Swords from Plowshares

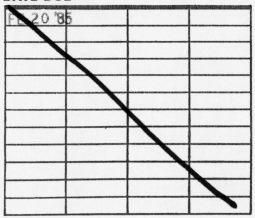

Swords from Plowshares

The Military Potential of Civilian Nuclear Energy

With a Foreword by Fred Charles Iklé

Albert Wohlstetter, Thomas A. Brown, Gregory Jones,
David C. McGarvey, Henry Rowen, Vince Taylor,
and Roberta Wohlstetter

The University of Chicago Press
Chicago and London

Library of Congress Cataloging in Publication Data

Main entry under title:

Swords from plowshares.

"A shorter version . . . originally appeared in
Minerva."

    Includes index.
    1.   Nuclear nonproliferation. 2.   Atomic
power—International control. 3.   Atomic weapons.
I.   Wohlstetter, Albert J.
JX1974.73.S96   327'.174   78–56373
ISBN 0–226–90477–6

The University of Chicago Press, Chicago 60637
The University of Chicago Press, Ltd., London

85   84   83   82   81   80   79   5 4 3 2 1
ISBN: 0–226–90477–6
LCN: 78–56373

Contents

Foreword

MANY nations now possess, although few have used, the materials and technology to build nuclear weapons. How rapidly and how far would this awesomely destructive potential have spread, had nuclear technology offered no peaceful applications? It is quite likely that such a situation would have made little difference to the four powers that followed the United States in acquiring a substantial nuclear arsenal—the Soviet Union, Great Britain, France, and China. But beyond these nations, the diffusion of the know-how and means to make "the bomb" would have been substantially slower. With the benefit of hindsight we can now clearly see that projects for peaceful applications of nuclear technology provided the essential expeditor, and in many cases the necessary cover, for gaining capabilities to make the bomb.

This book reveals how peaceful and military uses of nuclear technology have become intertwined. It shows the opportunities we missed and points out forcefully what still can be done—what must be done—to halt or at least to slow down the proliferation of this destructive power.

The research project on which this book is based was among several projects sponsored by the United States Arms Control and Disarmament Agency to give the U.S. government a better understanding of the dynamics and dangers of the nuclear spread and to help develop policies that could reduce these dangers. Rarely has scholarly research been so immediately influential for changing government policy. This study, far more than any others on this topic, revolutionized the thinking in the United States (and in other countries as well), leading the way to the radical new departure in U.S. nonproliferation policy that took place during the Ford administration. This new approach took full account of the essential technical and economic facts, primarily by seeking to establish a meaningful dividing line between peaceful applications of nuclear energy and activities undistinguishable from preparations for a nuclear arsenal.

The Carter administration's policy against nuclear proliferation, while adding certain new measures and introducing changes in tactics, essentially incorporated this "revolution." The findings reported in this book were later confirmed by a Ford Foundation–sponsored project (directed by the MITRE Corporation), a development which made the new approach to proliferation more widely accepted.

Some of the most essential findings of this project have meanwhile become accepted by groups which initially were violently opposed to these ideas—a true measure of success for policy research. Experts representing or closely associated with the nuclear industry now refer to the "consensus" that plutonium recycled for light water reactors is of dubious economic attraction. The opposite notion—that such recycling was essential for economic reasons—was long stubbornly held and has been responsible for much of the appalling diffusion in nuclear weapons material which occurred over the last decade. And initiatives are now being taken by proponents of nuclear power to change the plutonium fuel for the breeder reactor so as to make it unsuitable, or less suitable, for weapons production. This effort provides an explicit admission that plutonium does constitute a serious danger, an idea which had previously been denounced and ridiculed by the defenders of particular technologies.

Naturally, an intellectual effort that wrought such vast change could not escape bitter attacks from the vested interests that only reluctantly—and tacitly—came to admit that they had been wrong. And obviously, not all the findings of this project have become universally accepted. But there is no disagreement about the extraordinary influence of this research. A leading, highly knowledgeable spokesman for the nuclear industry said, in discussing the recent change in U.S. policy, that "the most significant single event . . . was the appearance in December 1975 of [the study] for the U.S. Arms Control and Disarmament Agency entitled, 'Moving toward Life in a Nuclear Armed Crowd?' "[1]

Misleading labels for technological projects can conceal important policy issues. "Project Plowshare" was the name given in the 1950s to the plan of using nuclear explosions for peaceful purposes. Regardless of whether or not this project would have been economically sound—and according to today's consensus in the United States it would not have been economically justified for all plausible applications—the labelling was plainly misleading. It made it easier, for example, for the Indians to pretend their nuclear explosion served only peaceful purposes. People think of swords that have been forged into plowshares as harmless because they cannot be used for war; but so-called "peaceful" nuclear explosions can destroy a city or other targets. Indeed, a basic weakness of some of the past policies has been the assumption that an application of nuclear energy will not be of military use simply because it has some civilian use.

Some egregious errors were committed in the past, both in the large thrust of overall policy and in the detailed but highly important technical decisions. Some of these errors were the result of oversight or stemmed from difficulties of prediction (so common in policy-making). Other errors can be traced to the wrenching emotional content of the subject matter. And still others—a chilling point—were clearly caused by malfeasance and deliberate suppression of facts.

That it is undesirable for atomic weapons to spread is an idea as old as the nuclear age. And it was recognized from the very outset that the peaceful applications of nuclear technology would therefore have to be restricted. At the beginning, of course, there was the American nuclear monopoly, offset only by some obligations to the United Kingdom and Canada, whose nationals and resources had contributed to the Manhattan project. Because of America's monopoly and unquestionable economic-

---

[1] Carl Walske, president of the Atomic Industrial Forum, in a speech entitled "Nuclear Power and Nuclear Proliferation," delivered in Geneva, September 27, 1977, to the international conference "Nuclear Power and the Public—A European-American Dialogue."

industrial predominance, it did not require much of an international consensus at that time to prohibit the transfer of nuclear technology and materials to other countries.

This, then, was the outlook in 1945. The ashes and agony of Hiroshima and Nagasaki provided painfully vivid proof of the destructive potential of atomic energy, while the peaceful uses of this new source of power were nothing but forecasts and speculations. It was therefore natural and easy for President Truman and the Canadian and British prime ministers to agree in November 1945 that there should be no disclosure of information, "even about the industrial applications" of atomic energy, until an international system of control was set up.

Some twenty years later these priorities had become completely reversed. In 1968, United States ambassador Arthur Goldberg told the United Nations in a speech in support of the Non-Proliferation Treaty that it would be an "unacceptable choice," indeed "unthinkable," to decide that the nonnuclear countries "must do without the benefits of this extremely promising energy source, nuclear power—simply because we lack an agreed means to safeguard that power for peace." The plowshares had now to be distributed throughout the world even if they could easily be forged into swords.

To be sure, there were to be agreements and safeguards against "diversion," and plutonium was widely supposed to be "denatured" in the course of its production in a power reactor so that forging a nuclear sword from it was presumed to be difficult if not impossible. But, in fact, the "denaturing" was a myth and, in practice, the agreements and safeguards came to be subordinated to the spread of civilian nuclear technology.

During the 1950s and 1960s, the United States was the principal supplier of nuclear assistance and also acted as the driving force in negotiations for the Statute of the International Atomic Energy Agency and the Non-Proliferation Treaty. Yet both these accords strongly tilt toward the potential beneficiaries of nuclear assistance by stipulating that safeguards must not hamper the flow of nuclear know-how and materials claimed to serve peaceful ends. Thus, article II of the Statute of the International Atomic Energy Agency states that the agency "shall seek to *accelerate* and enlarge the contribution of atomic energy to peace" and "shall ensure, *so far as it is able*" that its assistance is not used to further any military purpose (italics added). But if the agency is not "able" to ensure this, must it still *accelerate* the spread of nuclear technology? Similarly, the Non-Proliferation Treaty says (article III, 3) that the safeguards must "avoid hampering" international cooperation in the field of peaceful nuclear activities, and article IV has all the parties "undertake to facilitate . . . the fullest possible exchange of equipment, materials and scientific information for peaceful uses of nuclear energy." Article IV explicitly says that these uses of nuclear energy are to be "in conformity with Articles I and II." Indeed, it had to say that or it could hardly be called a nonproliferation treaty. But the practice moved in the opposite direction.

This curious evolution has a tangled history. By the mid-1950's, the world had come a long way from the aftermath of Hiroshima and Nagasaki, when secrecy and stringent controls on nuclear technology represented the approach that the public expected and that governments imposed. For the United States, there were not just two contending objectives to be sorted out (the containment of the atom's military potential, on the one hand, and the exploitation and worldwide diffusion of its peaceful uses, on the other); American nuclear and foreign policies were driven by a third overarching concern—the threats posed by the Soviet Union.

The very first American effort to prevent the accumulation of nuclear arsenals throughout the world—the Baruch proposal—was not unmindful of a future Soviet military threat. And it foundered because of Soviet opposition. At that point, the issue was simplified, at least for a few years. The approach to the overall conflict with Stalin's Russia and the approach to the risk of nuclear proliferation led to one and the same prescription: severely restrict access to and knowledge of this new technology. Keeping atomic know-how under tight secrecy meant keeping it from Russia.

In the early 1950s, however, as the Soviets successfully tested increasingly advanced nuclear weapons, it seemed reasonable to argue for a relaxation in the tight secrecy imposed by the United States government on American nuclear know-how. Three forces combined to press the Eisenhower administration for a greater release of nuclear data. First, the foreign policy argument: since the principal adversary apparently knew most of these secrets, it was in the United States' interest to share some of this information with allies and friends. Second, there was pressure for reducing secrecy about the effects and dangers of nuclear weapons: the public should not be kept in the dark about the horrors of nuclear warfare. The news media urged, and many of the scientists in the atomic weapons program supported, greater openness to alert the world public to the dangers of nuclear arms. In the fall of 1953, the Eisenhower administration began to prepare a program of disclosure, dubbed "Operation Candor," in response to these arguments.

But a third "lobby" now gained the upper hand. Nuclear scientists, government officials, and representatives of industry pressed for greater freedom to pursue the peaceful benefits of nuclear technology at home and abroad. The fact that the British seemed to move ahead of the United States in the development of nuclear power reactors provided an added reason. The idea of stressing the benefits of nuclear technology, instead of dwelling on the horrors, proved attractive to President Eisenhower and his advisers. Thus, in December 1953, the "Atoms for Peace" program was suddenly born.

To understand the first few years of this program, we have to keep in mind that the Eisenhower administration thought of it primarily as an instrument of foreign policy. The early tactics, therefore, were strongly influenced by the political competition with the Soviet Union. In the post-Stalin era, a U.S.–Soviet agreement for cultural and scientific exchanges had been reached. It was Washington's hope that this exchange program might lead to some opening-up of the closed Soviet society, and peaceful applications of nuclear technology seemed to offer promising ground for reaching Soviet scientists. At the same time, American largesse in the field of nuclear technology could be used as a tool to compete with the Soviet

Union for influence and prestige in many parts of the world. Surely, this must have been a large part of the reason why the United States gave 26 research reactors to foreign countries, including Indonesia, Thailand, South Vietnam, and Zaire.

We should recall the motives and mood of the times. For example, in 1955 a congressional delegation that had visited India reported that they were given "well founded testimony on the complete unwillingness of the Soviet Union to bring any real benefits and contribute any real scientific and engineering knowledge to the end that the people of India may enjoy the benefits of atomic energy." The congressional delegation favored giving India an assurance of a reliable U.S. supply of heavy water: "Such an assurance from us would constitute the kind of genuine cooperation from the United States that is needed to prove our good intentions."

At the same time, President Eisenhower's proposal for an international agency to control the distribution of nuclear technology and materials ran into opposition from the Soviet Union as well as from India and other countries. The United States, therefore, went ahead with bilateral agreements. The enthusiasm of some American officials for the transfer of American technology and materials was such that the recipients of these American gifts were accorded substantial bargaining power in whittling down the proposed safeguards. Our eager government officials somehow felt it was the United States that had to "prove its good intentions," not the recipient countries to whom we proposed to hand over free or at cut rates the most dangerous materials man had ever fashioned.

In the mid–1960s, as the United States and many other countries pressed for a nonproliferation treaty, the Soviet Union, after disagreement with the Chinese about the use of its assistance to develop weapons, increasingly became a supporter of international antiproliferation efforts. It emerged as a strong advocate of a nonproliferation treaty, although seeking to maximize the inhibitions imposed on U.S. nuclear cooperation with its allies; and it ceased to oppose and eventually supported control measures to prevent the diversion of nuclear materials. Indeed, since 1974, when the Ford administration launched important negotiations among the suppliers of nuclear technology to strengthen export controls, the Soviet Union has collaborated willingly, and with few exceptions constructively, in this new approach to controlling proliferation. Thus the American objective of containing the spread of nuclear arms had at last become disentangled from objectives and concerns regarding the Soviet Union. But by now, unfortunately, the commercial-economic interests in conflict with nonproliferation were all the more entrenched, particularly among some of our principal allies.

All policies for arms control have to cope with conflicting objectives; efforts against nuclear proliferation are not unique in this respect. In fact, compared with other areas of arms control, measures against proliferation have been surprisingly successful. In the late 1940s, it would have seemed highly optimistic to predict that thirty years later only five or six nations would have acquired nuclear arms and that over a hundred nations would commit themselves through an international treaty not to manufacture nuclear weapons. However, the most important element in this success appears to have been the guarantees provided by U.S. military backing of alliances, rather than the direct arms control measures such as the Non-Proliferation Treaty.

For the near future, this situation promises to remain stable, short of some major crisis that would provoke countries on the threshold of a nuclear capability to take that desperate step. Yet this very risk that the present system of restraint may buckle under stress is what renders the widespread accumulation of plutonium so dangerous. This book provides disturbing data on the impending "overhang" of countries with enough separable—and separated—plutonium at hand for quickly assembling a small nuclear force. This "overhang" is attributable, in part, to the American promotion of and subsidies for nuclear reactors abroad and, in particular, for the recycling of plutonium.

The fragility of the present restraints against nuclear proliferation is only partly to be blamed on the conflicting broad objectives in U.S. foreign policy. Much damage has been caused by mistaken approaches to technical and organizational issues. The principal culprit, it seems, was the Atoms for Peace program, and it was not so much the idea which caused harm as the way in which it was executed.

Particularly harmful has been the lack of understanding of the physical realities that would govern diversions of ostensibly peaceful nuclear activities to a weapons program. When the United States Congress was asked by the Eisenhower administration to revise the Atomic Energy Act so as to permit the transfer of know-how and materials, far too much assurance was drawn from the fact that information on design of weapons would continue to be kept secret and far too little stress was placed on the dangers of spreading technology and materials for the bombs themselves. In short, Congress was unaware of the extent to which it authorized the dispersion of capabilities to make weapons.

This misunderstanding was compounded by the persistence within the U.S. government of a specific piece of misinformation, namely the claim that the plutonium from power reactors was normally not suitable for making bombs. It is not quite clear why so many technically competent people helped to propagate this erroneous notion. It is clear, unhappily, that some used it deliberately to deceive their superiors as to the dangers of plutonium reprocessing. Some West European statesmen, it seems, were thus misled to underrate the danger of plutonium from power reactors. Professor Wohlstetter's inquiry into this matter (using data that had already been declassified) has helped greatly to put an end to this deception, once and for all.

The initial motives of this sordid episode may have been totally honorable. Many nuclear scientists and engineers yearned for a clearly peaceful domain, separated from the horrors of the bomb, in which their new technology could unfold and blossom. So they grasped at the idea of "denatured" plutonium. Given this desire for separating the two aspects of nuclear energy, however, it was strange—indeed irresponsible—that the engineers and scientists who developed the reactor technology during the 1950s and 1960s so badly neglected the technical factors that could help divide the peaceful from the destructive uses of the atom. Little effort was made to analyze this dividing line, let alone to strengthen it, until under President Ford American policy began to refocus on the technical aspects of proliferation.

Indeed, in some instances the dividing line was mindlessly weakened. For example, to reprocess spent reactor fuel, the designers chose the Purex method, which had been developed to produce especially pure plutonium—for what purpose? For making bombs. This Purex method, subsequently, was assiduously distributed by the U.S. Atomic Energy Commission throughout the world. Over 11,000 technical papers on plutonium technology were declassified in the proliferation avalanche released by the Atoms for Peace program. In addition, the constraints on the export of highly enriched uranium were progressively weakened.

How can this irresponsible behavior be explained? Why were there so few warning voices at that time? In part, the gross neglect of the dividing line between bombs and reactor fuel may have been caused by a myth which led the Atoms for Peace program astray: the myth that the job of guarding nuclear technology could all be turned over to an international organization, regardless of the type of materials and technical processes involved.

An insouciant lack of realism about the functioning and capabilities of the International Atomic Energy Agency has long bedeviled United States nuclear policy. American officials have often treated this agency like some closed magic box into which they can dump the problems that industry failed to solve at home. The agency made it easier for the exporters of nuclear technology in several countries to pretend that their practices were safe. Never mind that reprocessed plutonium could rapidly be manufactured into bombs—the agency would "safeguard" it. Never mind that highly enriched uranium was accumulating in large amounts in many countries—it was under agency "safeguards."

It is astonishing that the United States Congress tolerated this delegation of responsibility beyond its control and scrutiny. Indeed, the situation is worse than meets the eye. Article VII F of the Statute of the International Atomic Energy Agency enjoins the agency's staff from disclosing industrial secrets or other confidential information. This entirely proper restriction, however, is being used by American bureaucrats as an alibi for remaining ignorant about the inadequacies of the agency's safeguard system. Those who prefer to wash their hands of such inadequacies argue that they cannot be blamed for not knowing what happens to all the American nuclear materials and technology sent abroad: they are not supposed to know; it is "confidential" information of an international agency.

The U.S. Atomic Energy Agency (and its successors) completely lost track of the whereabouts of hundreds of kilograms of plutonium and highly enriched uranium. It took repeated prodding by the Arms Control and Disarmament Agency to start the job of compiling up-to-date data. Such data are, of course, part of the government's intelligence information, essential to protect the United States and its friends against possible misuse of nuclear materials, whether by national government or terrorist groups.

The execution of the Atoms for Peace program was further flawed by the sloppy drafting of many of the agreements for sending nuclear assistance abroad. United States influence on the disposition of nuclear materials was weakened, "peaceful" explosions were not always clearly prohibited, technology derived from U.S. exports escaped control, and so forth. To make matters worse, whenever a clear violation of the intent of U.S. assistance occurred—as in the case of India's nuclear explosion in 1974—some Washington bureaucrats would bend over backwards to interpret ambiguities (which they themselves may have written into the agreements) so as to exonerate the foreign government that had defied the intent of Congress. Here lies a lesson with broader implications for arms control agreements in general: ambiguities pose the double danger that they may be exploited by our adversaries and that those on our side with vested interests may cover up the facts.

Several misperceptions of the economic aspects of nuclear power also played a role in the spread of weapons capabilities. American reactor manufacturers failed to anticipate that by exporting their technology they would create other exporters that could take future sales away from them. Moreover, the reactor business acquired a false aura of profitability both in the United States and in Western Europe, because many of the costs were either borne by governments or obscured by the size and complexity of large corporations. Finally, early commitments by the reactor engineers to particular technologies, especially plutonium reprocessing, created strong vested interests in proving that their approach was profitable.

Vince Taylor and his collaborators uncovered the errors in the economic rationale for reusing plutonium as well as the questionable value of reprocessing for reducing the waste problem. Their research, now reported in this book, made a most critical contribution toward a more effective policy against proliferation. Once the economic case *for* plutonium was rendered doubtful, the cogent security argument *against* plutonium could gain the upper hand in Washington. Hence, in the fall of 1976, the Ford administration could announce a new nuclear energy policy which went a long way towards reestablishing the dividing line between bombs and reactors.[2]

Of course, the opportunities lost over the last twenty years could not be fully retrieved. The Carter administration initially sought to strengthen the new policy thrust, but countervailing forces here and abroad have partially succeeded in blunting it. As of this time, the outcome is in doubt.

This book—perhaps surprising some readers—is not opposed to peaceful uses of nuclear energy. It shows how the uses of this new energy can profitably expand in a way that is significantly separable from preparation for a weapons arsenal. By contrast, if we kept moving towards "life in a nuclear armed crowd," the nuclear industry would find itself far more embattled and restricted than today. So, indeed, would we all. Deterrence would become even more dangerous and delicate a strategy. And the need to cope with the multiplying risks of nuclear disasters would jeopardize our freedoms.

FRED CHARLES IKLÉ

[2] See the publications of the U.S. Arms Control and Disarmament Agency: *Nuclear Policy* (November 1976) and *Peaceful Nuclear Power Versus Nuclear Bombs: Maintaining the Dividing Line* (Publication 91, December 1976).

I

THE TRENDS AND THE QUESTIONS THEY POSE

> " Whatever else hospitals do, they should not spread disease."
> Florence Nightingale

THE analysis of trends in the spread of nuclear technology which follows is intended to raise more questions than it will answer. For good reason. In this complicated and fateful area of policy, very often actions taken for the best of ends, perversely defeat them. If our policies are to cope with the spread of military nuclear technology rather than encourage it, it is essential that they be more than symbolic and well-intended, more than " allusive and sentimental "—as Robert Oppenheimer called " atoms for peace ". They need to be concrete and aimed precisely at the problems posed by changes in the real world. Otherwise, like the nineteenth-century hospitals Florence Nightingale referred to, they are as likely to spread the disease as to cure it. But then, we need to understand underlying causes and effects. This study is deliberately a preface to policy.

We begin by stating summarily and plainly some central points.

Nearing the Bomb without Breaking Promises not to Make It

Many countries, including many agreeing not to make bombs, will come very close to it without precisely violating their agreement. They will get the fissile material—or the means to produce the material—bringing them very close to the manufacture of bombs, from supplying countries which have promised not " in any way to assist " countries without weapons to obtain them. And, like the importers, these supplying countries also can say that they are not violating their promises.

It has been understood from the outset of the nuclear age that designing a bomb and getting the non-nuclear components are much easier than getting fissile material in high enough concentrations for an explosive. Research on the design of bombs and testing of non-nuclear bomb components are not prevented by agreement, and can proceed in parallel with the accumulation of fissile material. Obtaining fissile uranium (in particular uranium-235) or fissile plutonium (especially plutonium-239) concentrated enough to need no isotope separation [1] and only a modest amount of chemical separation are then the main obstacles in the path to a nuclear bomb.

By 1985, according to their plans in 1975, nearly 40 countries will have enough chemically separable plutonium for a few bombs in the spent fuel

---

[1] Isotopes of the same heavy element, such as uranium-235 and uranium-238, undergo the same chemical reactions at almost the same reaction rates and therefore cannot be separated by any known conventional chemical means, but so far only by an expensive, difficult and time-consuming physical process which exploits slight differences in atomic mass. The fissile isotopes are those which are readily fissionable by slow or thermal neutrons as well as fast neutrons. The actinides include some isotopes such as neptunium-237 with an even number of neutrons which, even though they are not fissile, have finite critical masses in fast neutron systems and so may form material for explosives. However, these are likely for some time to come to be much scarcer than the fissile isotopes.

produced by their electrical power reactors. About half these countries have been planning a capacity by then to separate at least that much plutonium from the spent fuel.

A few years later, if the recycling of plutonium has become general, many governments will come significantly closer to obtaining chemically separated plutonium metal, even if they do not themselves separate plutonium from spent fuel. Those which manufacture plutonium dioxide fuel rods will have plutonium easily available in their inventories at the start and during the process. If they neither separate plutonium nor manufacture their own plutonium fuel rods, nonetheless about 25 or more countries will have very large quantities of plutonium—enough for from 50 to 1,400 bombs—in fresh fuel which has not been irradiated.

Reactors used for the propulsion of ships, some proposed forms of electrical power reactors, such as the high temperature gas cooled reactor (HTGR), and a good many research reactors now operating, use as fuel highly enriched uranium of the sort needed for weapons. Many " research " reactors, like the CIRUS reactor given by Canada to India, can produce sizeable quantities of plutonium for explosives. Each of these is a potential source of material for weapons, obtainable without violating agreements. The explosive material these sources might provide will be dwarfed, however, by the plutonium output of electrical power reactors owned by an increasing number of countries.

Critical experiments for research on fast breeder reactors may call for stocks of several hundred kilograms of plutonium, which are enough for about 50 bombs or more, and comparable quantities of highly enriched uranium, all almost immediately available for use as an explosive. Such experiments are not in general excluded by agreement.

All this then can happen without violating any agreements—at least any clearly understood, unambiguous agreements: a growing and legitimate—but Damoclean—" overhang " of countries increasingly near the making of bombs. This " overhang " of countries is additional to those which might acquire the bomb by an overt military programme which they have not foresworn. Or by cheating. Or by shopping, as Libya has tried, for a finished bomb.

For this major problem of a growing, legitimate " overhang ", safeguards against diversion will become increasingly irrelevant. The " overhang " requires no diversion, and therefore no violation of safeguards. Suppliers of plants which separate fissile material—by chemically reprocessing plutonium or enriching uranium in the fissile isotope, uranium-235 —frequently say, correctly, that the facilities they propose to supply will be subject to arrangements for international inspection at least as stringent as any provided for currently. This may suggest that such " safeguards " would make the safeguarded activities " safe ". They would not: not if the activities involved result in stocks of separated plutonium or highly enriched uranium in the hands of governments which do not now have nuclear weapons.

The bilateral and trilateral agreements on nuclear cooperation into which the United States has entered generally leave title to the plutonium in the countries operating the reactors, subject only to " safeguards ". But the stocks of material so acquired can grow legitimately while being closely watched by an inspector.

The distinction between " safe "—civilian—activities and " dangerous " —military—activities is becoming increasingly ambiguous. This ambiguity will confuse and reduce the clarity of a warning that bombs are about to be made, or that they can quickly be made. Such ambiguities also weaken sanctions: sanctions impose some costs on suppliers as well as on recipients, and the less clear-cut the violation, the weaker the resolve of the supplier.

Sophisticated proponents of a system of safeguards stress that it is primarily a kind of " early warning " alarm bell. However, as things stand, the system of safeguards of the International Atomic Energy Agency (IAEA), which is designed to detect diversion, may actually muffle signs of the critical changes taking place without diversion.

The early warning which has always been implicit in the idea of safeguards would permit a government to respond appropriately and in time, even if no collective sanctions have been defined. But the IAEA rules, designed to protect commercial and industrial secrets, obscure the information relevant for governmental decisions on security.

The IAEA treats as, " safeguards confidential ", information reported to it by the countries it monitors regarding actual quantities of nuclear materials and their physical and chemical states. It does not show this critical information to any other government, much less to the public. As a result, not much is known in or out of governments, in any regular way, about the distribution by country of stocks of fissile material.[2]

The Nuclear Weapons Revolution for the Less Developed World?

The countries moving towards acquisition of nuclear explosives tend now to be the small or less developed ones, especially those outside the Soviet orbit, not—as was once expected—the most advanced industrial powers. The situation may be like that of the Marxist revolution: predicted for the advanced bourgeois countries, it came in the backward ones. This development, among other things, may complicate the problem of slowing the spread of bombs by merging with the demand by the less developed countries for equality with the advanced industrial countries, many of which are also exporters of nuclear material.

Unofficial nuclear terror: The problem of " sub-national " or " transnational " nuclear terror seems most acute in those countries which have the most immediate prospects for acquiring nuclear weapons. Nuclear terror, while possible in the United States, has received, because of its sensational character, an excessive amount of attention there. The United States government, the Congress and the press gave it more notice than the problem of proliferation, until the recent change in American policy deferring commitment to plutonium. An unofficial nuclear terror may

2 Chapter III discusses the current American attempt to remedy the situation (pp. 65–70 and fn. 47).

become a genuinely serious problem as a result of the approach by many governments to the capability of making nuclear weapons. Some of the countries which may soon acquire nuclear weapons are politically unstable and much more liable to sudden threats of mass destruction from dissident factions.

The bomb as a substitute for alliance: The most likely countries to decide to acquire nuclear weapons appear to be " unaligned " by choice or to be outcasts, dropouts, or fading members of alliances, especially the system of alliances of the United States: South Korea, shaken by the collapse in Southeast Asia; Taiwan by that and by the American rapprochement with the People's Republic of China; Pakistan, feeling abandoned in its conflict with India by the United States and the United Kingdom, and now facing an India which has shown that it can make and detonate nuclear explosives; Iran, now following a policy of " independent nationalism " in the Middle East and interested in keeping external powers out of the Persian Gulf; Brazil and Argentina asserting independence in the Western hemisphere; Spain so far excluded from NATO; and South Africa which is " odd-man-out " of all alliance systems.

In the same way, the few communist countries which might move in this direction are either outside the Soviet orbit, as in the case of Yugoslavia, or moving away from it in foreign policy, like Rumania.

The Economics of Nuclear Power in the Developing Countries

For most developing countries, where civilian programmes are moving towards a military capability, the civilian economics look particularly poor: nuclear electrical power is capital-intensive and is likely to be economic, if at all, only on a very large scale; it requires continuing specialised and highly sophisticated operation, maintenance and resupply. Less developed countries are in general short of capital, need electrical power generated in small units fitting their small electric grids, and have a restricted base of skilled human capital relative to their needs.

The increase in oil prices by the Organisation of Petroleum Exporting Countries (OPEC) intensified the scarcity of capital and foreign exchange in those less developed countries which are not members of OPEC. Soon after the October war had led to a jump in fuel prices, the IAEA predicted that nuclear electrical power would be economic in most developing countries, compared with fossil fuels including indigenous and imported coal. These predictions have had to be revised. On the other hand, the less developed countries in OPEC which have suddenly acquired a great deal of capital have very large supplies of fossil fuel and in many cases small demands for energy. An extensive shift to nuclear electricity would not only be uneconomic, it would greatly increase their dependence for energy on equipment, materials and technological aid from outside powers. Even the oil-importing less developed countries might use new capital better to reduce their dependence on fossil imports by exploration and fuller exploitation of their known reserves. It has been estimated that for these countries an investment of $60 billion for such purposes could by 1985 reduce their total imports of oil to zero. Given their limited ability to service very large current debts, and the very long delay before nuclear electrical power could constitute a significant fraction of net energy consumed, investment to develop their own mineral resources may be a more promising way to reduce dependence on imports of fossil fuel and other products and services related to the supply of energy. For example, Brazil's very ambitious programme, which includes the import of a small enrichment plant from Germany, will still leave it dependent, among other things, on the import of slightly enriched uranium.

Investment in nuclear energy is a poor choice among alternatives for the economic development of less developed countries. It diverts capital from more productive uses. Instead of speeding economic development and slowing the spread of military technology, as we had hoped for decades, the subsidised transfer of nuclear technology has slowed development and may speed the spread.

If the economics of nuclear power in its safer forms are dubious in the developing world, this is even more the case for the economics of early commitment to forms of nuclear energy which are most dangerous from the standpoint of spreading nuclear explosive material. Reprocessing plants of optimal size for recycling plutonium in light water reactors (LWRs), for example, require spent uranium in amounts which would not plausibly be available for many decades even from the most ambitious reactor programme of any developing country. Plutonium breeders are a long way off and are expected to be even more capital intensive than light water burner reactors. They are also likely to require greater sophistication for their operation and maintenance; and their optimal sizes are expected to be about twice as large as current LWRs which themselves are in general too big for the small electric grids in the developing world. (European and American research and development have centred on breeder reactors with capacities of about 2,000 megawatts each.)

If considerations of conservation and "energy independence" are unlikely to compel an early commitment to these more dangerous forms of nuclear activity in developing countries, neither do the necessities of managing and finally disposing of radioactive waste. Reprocessing spent uranium creates new forms of radioactive waste and large volumes of low and intermediate level waste as well as high level waste in liquid form, and increases the difficulties and dangers of interim waste management, which have so far not been handled well in the most advanced nuclear countries. Final disposal in geologically isolated areas may be troublesome in small countries, and particularly in the seismic regions characteristic of several countries in Asia and the Middle East which have exhibited an interest in reprocessing spent uranium fuel and the use of plutonium.

However, for the advanced industrial countries also considerations of economics, of conservation and of waste disposal do not compel a commitment now to the use of plutonium or highly enriched uranium as fuel.

The Poor Economics of "Dangerous Nuclear Materials" in the Industrial Countries

No reactor using highly concentrated fissile material as fuel, such as plutonium or uranium highly enriched in uranium-235, has yet had any significant commercial use. But it has been long anticipated that such fuels would become widespread and that it would be necessary to transform the relatively abundant but inert isotope uranium-238 into fissile plutonium, to supplement or replace the active isotope uranium-235 which occurs in nature as less than 1 per cent. of yellow cake (U_3O_8). The economically usable uranium-235 was thought to be not only rare compared to uranium-238, but to be very scarce on any terms. In fact, the idea that a plutonium breeder was essential if nuclear power was to have any significant future originated during the Manhattan Project, when in 1944 it was estimated that the economically recoverable reserves of uranium yellow cake would amount to only 5,000 tons in the United States and only 20,000 tons worldwide.[3] The entire world supply, so estimated, would have been enough to generate for only the equivalent of a year and a half electrical energy in the quantity which the United States was using at the time. Moreover, the official American estimate made in 1948 halved the earlier figure.

The belief in the necessity of breeding plutonium persisted long after the United States produced three orders of magnitude more uranium than it had estimated in 1948, and long after reserves and " potential reserves " available at reasonable economic cost had multiplied by an even larger factor. Meanwhile, the economic prospects for forms of nuclear electrical power which would spread nuclear explosive material, such as plutonium, appear considerably less attractive or much more remote and uncertain. Programmes for such nuclear processes need to be looked at with particular sobriety and scepticism. It would be worth a substantial economic sacrifice to avoid the large political costs in allowing the development of a world of many states armed with nuclear weapons. Some such sacrifice might conceivably be involved in restricting nuclear electrical power to less dangerous forms which minimise access to readily fissionable material. However, economic evidence indicates that such extra costs are not at all likely to be large in relation to total electrical power costs; and they may be negative.[4] We may save money as well as potential trouble by forgoing some of the plainly dangerous activities now contemplated.

The expected costs for power from light water reactors with plutonium recycling have increased drastically compared to such power without recycling. The estimated costs for separating plutonium have multiplied over tenfold in little more than a decade. They have, in fact, even allowing for inflation in the general price level, jumped by a factor of nearly seven in two years, between the estimate by the United States Atomic Energy Commission (USAEC) of $30 per kilogram in 1974 and the estimate of $280 in mid-1976 by the agency which succeeded it, Energy Research and Development Administration (ERDA).[5] In the negotiations

[3] Wohlstetter, Albert, *The Spread of Military and Civilian Nuclear Energy: Predictions, Premises and Policies* (Los Angeles: Pan Heuristics E–1, 1976), p. 15 ff.

[4] See Taylor, Vince, *The Economics of Plutonium and Uranium* (Los Angeles: Pan Heuristics, 30 April, 1977), and Chapter IV below.

[5] United States Atomic Energy Commission, *General Environmental Statement on Mixed Oxides (GESMO)* (draft), WASH-1327 (Washington, D.C.: U.S. Government Printing Office, August 1974), Vol. IV, pp. xi–41. For later estimates, see Chapter III below.

during 1977 between COGEMA of France and DKW of West Germany the price of reprocessing a kilogram of spent uranium dioxide has been reported to be about $500 in constant 1977 dollars.[6] Evidence presented by British Nuclear Fuels Limited (BNFL) at hearings on the expansion of its plutonium separation plant at Windscale indicate a cost of £260 ($460) per kilogram in mid-1977 pounds.[7] On lower estimates reached by Pan Heuristics, the future costs of plutonium fuel will nonetheless exceed those of fresh uranium.

On the latter calculations, extracting plutonium from the spent uranium of light water reactors and fabricating the plutonium into fuel rods will be more expensive than the fresh uranium fuel rods they would replace: 50 per cent. more in a large plant processing daily 5 metric tonnes in uranium content (MTU) of spent uranium reprocessed in the United States; 80 per cent. more in such a plant in a multinational nuclear centre; 400 per cent. more in a plant processing one MTU daily—a very large plant for a less developed country, since it could service 10 reactors of 1,000 MWe each. For plutonium extracted from heavy water reactors of the sort purchased from Canada by Argentina, Pakistan and India, costs might double again. Recycling fissile material from heavy water reactors is even less economic than from light water reactors since the spent fuel from heavy water reactors contains plutonium in more dilute form and uranium with much lower content of uranium-235 than natural uranium. Recycling in light or heavy water burner reactors, as distinct from breeder reactors, in short, seems likely to lose rather than save money.

Advocates of plutonium disagree [8] and of course both their estimates and those of Pan Heuristics are quite uncertain. However, the area of agreement is more important and less in doubt. Both sides recognise that, even on the most optimistic estimates, recycling of plutonium in light water reactors can make only a slight difference in the costs of a delivered kilowatt hour. Fuel cycle costs are only about one tenth of delivered costs of electricity, and recycling can displace only a fraction of the fuel cycle costs; the total saving would be at most only 1 or 2 per cent., even if plutonium separation were costless.

However, given that the main issue today, which has been raised by the initiatives of Presidents Ford and Carter, is the limited one of whether to defer the decision or commit ourselves now to plutonium fuel, the area of disagreement should be smaller still. On the analysis of the *Final Generic Environmental Statement on Mixed Oxides* in mid-1976,[9] advocates of an immediate commitment on the staff of the United States Nuclear Regulatory Commission (NRC) estimated that an eight-year delay

6 *Nucleonics Week*, XVIII, 50 (15 December, 1977), p. 6.

7 " A Cost Assessment of Alternative Methods of Dealing with Irradiated Oxide Fuel ", BNFL, Document 232, submitted at the Public Local Inquiry into an Application by BNFL for Permission to Establish a Plant for Reprocessing Irradiated Oxide Fuels . . . at Windscale, June 1977, p. 3 (typescript). The evidence is not detailed enough to fix precisely the implicit cost per kilogram.

8 For details, see the comparison of five recent analyses in Taylor, V., *op. cit.*, Appendix A; and also " Is Plutonium Really Necessary? " (Los Angeles: Pan Heuristics, July 1976).

9 United States Nuclear Regulatory Commission, *Final Generic Environmental Impact Statement on the Use of Recycled Plutonium in Mixed Oxide Fuel in Light Water Cooled Reactors*, NUREG–0002 (Washington, D.C.: Office of Nuclear Safety and Safeguards, August 1976), Vol. IV, p. xi–59.

would cost \$74 million in discounted 1975 dollars over a 25-year period; and a delay of 13 years would cost \$300 million. These sums are very small compared to the discounted sums which would be invested in the nuclear fuel cycle and in reactors during that period. The advocates' own reckoning implies that the cost of an eight-year delay would be measured in hundredths of 1 per cent. of the cost of a delivered kilowatt hour of electricity. Even a delay of 13 years would add less than one tenth of 1 per cent. to delivered kilowatt hour costs. In the light of the long and disastrous history of failures in predicting nuclear costs, demand and supply, it would take a heroic faith to assign any reality, much less an importance, to such small differences.

On the other hand, since investment expenditures are made early and benefits received much later, premature commitment can mean a very large financial penalty in the event of failure. That penalty may be passed to foreign or domestic utilities or borne by the manufacturer, but ultimately it will be borne by taxpayers or users of electricity. As we have seen, such economic arguments apply even more forcefully to the Third World countries which contemplate reprocessing, since they would experience much higher costs through the diseconomies of small scale. The political and military penalties imposed by early commitment to separating plutonium are even larger and some may be substantially irreversible.

Conservation and the Dangerous Processes

The benefits of conservation must be related to the economics. The supply of uranium worth finding and extracting is, of course, a function of expected price. Whether we extract plutonium from spent uranium fuel or mine fresh uranium ore depends on future relative costs that are very uncertain. Our estimates show that plutonium fuel would cost more than it is worth. But the fact that the mining and milling costs of uranium oxide are only half the fuel cycle costs and thus about 5 per cent. of the delivered kilowatt hour costs, suggests that the price of uranium yellow cake, or U_3O_8, could rise by a large factor without prohibitive effects on the costs of electricity. Plutonium in light water reactors can displace at most a modest fraction of uranium fuel, which will vary among other things with the assumed rate of growth of nuclear electrical power. The more rapid the rate of growth and the consequent demand for nuclear fuel, the smaller the fraction of fresh uranium fuel it can displace. Thus in various projections it has been estimated to save between 10 and 25 per cent. An estimate made in December 1975 by the Organisation for Economic Cooperation and Development, when the expected demand for uranium was still greatly inflated, indicated that, for the entire non-communist world, the cumulative savings from plutonium recycle until the year 2000 might amount to 9 months' supply of uranium at that date. Together with the recovered uranium, the recovered plutonium might reduce requirements for fresh uranium by about 10 per cent.[10] An estimate of the same date by the Edison Electric Institute

[10] Recovered uranium is contaminated with U-236 and is, therefore, equivalent to only a little more than one half the amount of uncontaminated uranium. We have taken this

(EEI) suggested a cumulative saving by the year 2000 in the non-communist world of about 13 per cent., or slightly more than OECD's. For the United States, the EEI numbers indicated savings in uranium of about 10 per cent. by 1990 and 16 per cent. by the year 2000.

Assuming less inflated rates of growth in nuclear power, the fissile material recovered from spent uranium could displace a somewhat larger percentage of the fresh uranium, but the supply of fresh uranium being larger in relation to demand, the impression of a "uranium crisis" demanding the early use of plutonium fuel is even harder to sustain.

But the issue is a narrower one if we are considering whether to commit ourselves to plutonium fuel. How much would a delay of 5 or 10 years in this commitment offset possible savings in uranium? The answer is very little. If the spent uranium fuel is placed in retrievable storage, as is contemplated, nearly all the fissile plutonium would be there ready for use if desired, since some 85 per cent. of it is plutonium-239 with a half-life of about 24,000 years. The 15 per cent. is constituted by plutonium-241 which has a half life of 13 years. As a result, after five years, 96 per cent., and after 10 years, 94 per cent. of the total fissile plutonium would remain. That would mean that if previously the plutonium were going to displace 20 per cent. of the uranium, it would then be able to displace only about 19 per cent. On the other hand, the spent uranium would have cooled meanwhile, making it somewhat less expensive to reprocess.

Conservation and the breeder: Stocks of plutonium or other highly concentrated fissile material might be used for initial loadings of fast breeders. The breeder has from the start been regarded as the ultimate method of conserving uranium. It has provided a principal rationale for stockpiling separated plutonium since the 1960s, and in some cases even as early as the 1950s.[11] India, for example, decided to get a plutonium producing reactor and a separation plant in 1956, long before it had any power reactor and at a time when even its long-range programme called for the use of plutonium fuel only after the advent of thorium breeders or near breeders.[12] The accumulation of stocks for the initial loading of the breeder is not a very convincing economic argument for expanding commitments to separate plutonium now. The Indian example is in fact a reduction to absurdity of that argument.

It should be understood that the argument has to do not with the stocking of plutonium, but with the time when separating it might be justified. For the breeder as for the light water reactor, the predominant plutonium-239 would remain undiminished in storage, whether separated

penalty factor into account in this and following estimates of resource savings from recycling of uranium.

[11] Richard J. Barber Associates, *LDC Nuclear Power Prospects, 1975–1990: Commercial, Economic and Security Implications*, ERDA 52–UC–2 (Washington, D.C.: 1975). Section III, p. 4.

[12] Wohlstetter, Roberta, "*The Buddha Smiles*": *Absent-Minded Peaceful Aid and the Indian Bomb*, Report E-3 (Los Angeles: Pan Heuristics, April 1977).

or not. And the breeder would be at least equally affected by the decay of the small fraction of the total consisting in plutonium-241. If the plutonium were separated early for much later use in the breeder, not only would the cost of separation be higher because of the more intense radioactivity, but the separated plutonium itself would be more hazardous and might involve more costly handling, since plutonium-241 decays into americium-241, an intense γ-emitter. This is to say nothing of the advantage which might be gained by waiting for the results of further research into techniques of separation and waste disposition; nor of the advantage which might accrue from reserving the decision on whether or not to separate plutonium at all.

The last point is essential. It is uncertain when, if ever, the breeder will become viable commercially. An investment in the 1950s for the ultimate purpose of loading thorium breeders (which, it is clear now, are unlikely to have much commercial significance before the second quarter of the twenty-first century, even in advanced industrial countries), meant immobilising extremely scarce resources in India. These resources could have been put to use to advance economic growth and therefore to increase the resources available to later generations. On the other hand, from the standpoint of proliferation, the dangers are precisely illustrated. The premature investment in Indian facilities for producing and separating plutonium for a breeder made it easier for the Indians to change their minds about nuclear explosives and to alter the purposes of their facilities in response to their defeat by China in 1962, to the Chinese nuclear explosion in 1964, and to the Chinese nuclear programme which followed. As many Indian officials then said, their civilian nuclear programme had greatly reduced the additional costs of making their own nuclear explosives.

Whether the fast plutonium breeder will ever be an economic way to produce fissile material depends not only on many technical developments, but on the future capital costs of various breeder and burner reactors; the processing costs of the breeder's intensely irradiated fuel; the future price of fresh uranium; the growth in demand for electricity; the reserve electric generating capacity, and many other variables. Even if it should become economic, still more uncertain are the dates at which it would become so, and the realistic rate and extent at which it might grow in commercial significance, without risk of losing enormous investments. The immense uncertainties are confirmed by the poor record of predictions by the United States and United Kingdom in the field of the spread of civilian and military nuclear energy.[13] They all suggest the importance of strategies of sequential decision which avoid a premature commitment and proceed to commitment as uncertainties are progressively resolved.

Reprocessing and the disposal of radioactive waste: Because from the start it had been assumed that the use of plutonium fuel would be necessary for the future of nuclear electrical power, it is only recently that the final disposition of unreprocessed radioactive waste has received

---

[13] See Wohlstetter, Albert, " Statement of Submission." 5 September, 1977, to the Public Inquiry at Windscale, *op. cit.*

attention. In the United Kingdom, the Federal Republic of Germany and Japan the official view continues to be that reprocessing is essential for the permanent disposal of spent uranium fuel, even if it is not economic for the recovery of fissile material for light water reactors. It has frequently been claimed that, compared to spent uranium fuel, the volume of waste requiring geological isolation and remote handling is greatly reduced by the separation of plutonium in the course of reprocessing, and by compacting and vitrification. It is also sometimes asserted that the spent uranium fuel will need reprocessing with little delay because it will deteriorate and become a danger during storage in water pools.

In the United States, Canada and Sweden, however, professional opinion of both advocates and opponents of a commitment to plutonium now holds that the permanent disposal of spent uranium dioxide fuel from heavy and light water reactors may be accomplished with or without reprocessing and with no substantial difference in risk.[14] In the case of final disposal, the primary barrier contemplated between the radioactive waste and man would be geological isolation in some formation which has been stable for millions of years such as bedded salt or granite, rather than in a man-made structure. A serious breach of containment, if the site is properly selected, is judged to be extremely improbable. For example, the repository being considered in New Mexico, once sealed, could not be breached by the surface burst of a 50-megaton nuclear weapon.

In the very improbable event that the geological barrier were to be breached, the reduced amount of plutonium in the recycled waste from light water reactors might mean slightly diminished long-term risks in final disposal of the waste. However, it would do so at the expense of an increase in the formation of americium-243 and curium-244, and a consequent increase in the more worrying immediate risks of managing wastes before final disposal.

Chancellor Schmidt has expressed concern about the availability of space for final disposition of radioactive waste in smaller countries. However, reprocessing and the fabrication and irradiation of plutonium fuel create more and new forms of waste which require remote handling and geological isolation, and which even after compaction have a larger volume and heat content than the spent uranium fuel itself. Since the space occupied in salt beds is likely to be determined primarily by the initial heat content of the waste, the total space occupied by reprocessed waste is likely to be larger than if the spent uranium fuel were placed directly in the repository without reprocessing. This may apply with

---

[14] Eric Svenke, a member of the Swedish Government Committee on Radioactive Waste, summarised that committee's findings: " If reprocessing and reuse is not to be implemented, the commission is convinced that a safe terminal storage of spent fuel can be developed as well." Typescript of lecture at IAEA, Vienna, June 1976, p. 7. Similarly the *Final Gesmo* by the U.S. NRC found no clear preference between disposal of reprocessed, recycled and vitrified wastes, and the direct disposal of unreprocessed spent uranium dioxide fuel, *op. cit.*, Vol. I, p. S–30. See also Hebel, L. Charles, *et. al*, " Report to the American Physical Society by the Study Group on Nuclear Fuel Cycles and Waste Management " (typescript) October 1977, p. 5: " The decision to reprocess nuclear fuel does not depend significantly on waste management considerations."

particular force to the bedded salt repository in the Federal Republic, since its standards require geological isolation of even low level wastes.[15]

As for the contention that there will be a rapid deterioration of spent fuel rods during storage in water pools immediately after removal from the reactor and that this prospect compels reprocessing, it is not supported by studies of the behaviour to date of the fuels from heavy and light water reactors which are now standard.[16] There are special problems in some fuels of older design, such as the British magnox fuel, and in the fuel from the British advanced gas cooled reactors (AGR). They may need special handling, although not—as Sir Brian Flowers [17] has pointed out—necessarily reprocessing. However, there are few places outside the United Kingdom where these fuels have been used.

In short, from the standpoint of economics, the conservation of resources, or the management and final disposal of waste, there is no persuasive evidence—though it is often repeated at the expense of some confusion in policies aimed to prevent the proliferation of nuclear weapons—that plutonium is essential for the future of nuclear electrical power. And most important, there is no convincing argument for deciding now the long-term issues.

The Extremes and the Genuine Issue Today

The current public debate unfortunately tends to pit extremes against each other. On the one hand are the advocates of stopping and dismantling all nuclear electrical power plants as equally injurious to the environment and physical security; on the other are those who appear to defend all forms of nuclear energy as vital to the growth of advanced industrial societies and as critical for the economic development of poorer countries.

But the extreme alternatives are not in fact very interesting. It is plain, for example, that we will for a very long time use fossil fuels as well as nuclear fuels.

In the long term we shall want to develop a variety of alternatives for the safe transition to an era of abundant or renewable sources of energy. Among the nuclear choices, we need to develop and evaluate a great many possibilities which avoid unrestricted access to slightly radioactive, easily separated fissile material. We shall have to compare systems which would use natural or slightly enriched uranium only once in a reactor, from which the fissile material in the fresh fuel could be concentrated only by a difficult process of isotopic separation and in which the spent fuel containing fissile material separable by chemical means would be returned to the state with nuclear weapons which supplied it or would be subjected to strict control, perhaps in an international centre. This sort of system might be made universal, that is, used in states with and without nuclear weapons. But even such a system may imply at the same

[15] Hagen, Manfred, *et al*, *Atomwirtschaft* (July 1976), p. 338.

[16] Johnson, A. B., Jr., *Behaviour of Spent Nuclear Fuel in Water Pool Storage*, BPNL–2256/UC–70 (Battelle Pacific Northwest Laboratories, September 1977).

[17] " Record of the Sunningdale Seminar on Nuclear Proliferation ", Sunningdale, England, 13–14 May, 1977, p. 10 (typescript). See also the remarks of Dr. H. Dunster, deputy director, Health and Safety Executive, United Kingdom, Department of Energy, on p. 10.

time some restraints on the export of new technologies—for example, laser enrichment—making isotopic separation less difficult. Although these dangerous technologies might be confined to states with nuclear weapons or to international centres, the slightly enriched uranium they produced would be distributed.

Analogous arrangements might be appropriate for thorium breeders, and the fuels slightly enriched in uranium-233 which they might make economic. But in any case the technologies should be chosen together with the institutions in which they would be embedded. Such distinctions have been implicit in American export policies which in the past encouraged the sale of slightly enriched uranium, while prohibiting the export of the gaseous diffusion plants which enriched the fuel. However, over time the coherence of such policies has been eroded, and we need not only new technical alternatives, but a new systematic policy which will consider both the economics and the dangers of a branching process which might rapidly spread the ability to make bombs.

Many alternatives have been suggested recently, which might achieve comparable safety and use uranium more efficiently than the current light water reactors. But it is unlikely that any of the alternative systems will be perfected or will emerge as definitely preferable in the near future. The essential point, however, is that the immediate issue does not involve commitment now for the long term. The nuclear industry has in fact suffered from a chronic state of premature commitment and the chief political differences today lie between those who insist that we must commit ourselves now to plutonium and those who say that we can wait.

The Branching Process of Spread

If the ability to make or come close to making nuclear explosives spreads among less developed countries, the spread may at some later date alter the decisions of some advanced industrial countries which have so far deliberately forgone the development of an independent force of nuclear weapons. Advanced industrial countries like the Federal Republic of Germany and Japan have been both protected and constrained by their alliances. Alliances themselves may be weakened by the spread. (Nuclear weapons have long been promoted as a substitute for alliance.) If so, as the taboo weakens, some advanced countries may follow the less developed ones.

There are several branching processes which may increase the gathering instability we have described. The export of nuclear materials, equipment and services not only creates users of nuclear electrical power with the by-product of fissile material, it also creates other exporters. So Westinghouse created Framatome, General Electric created Kraftwerk Union, and so on. Now it seems India may be in the business of nuclear exports.

It is possible to trace serial connections between the Chinese and the Indian explosions on the one hand, and the public and private decisions to edge towards the making of a nuclear explosive in other countries. In fact, although the Indian explosion followed the Chinese by nearly 10 years, study of the development of the Indian programme plainly shows the link.

Case studies of other Asian and Middle Eastern countries show similar links with the Indian explosion.

These branching processes are, however, more complex than the exponential physical and biological processes which have suggested the standard metaphors of proliferation. They are not automatic, but depend on a complex set of political, military and economic conditions. Nonetheless these accumulating changes point to serious instabilities in the processes of decisions both to acquire and to use nuclear weapons. They add up to the prospect of a much more disorderly world.

Fatalism and Acceptance of the Spread

A feeling of fatalism on this subject is growing. A few years ago two of the ablest students of the proliferation of nuclear weapons commented that two views seemed to be emerging: one held that there was no problem —the Non-Proliferation Treaty would take care of it. The other held that there was no solution—nothing could be done about it. Today many seem to hold both these positions. They base the more comfortable position, namely that there is no problem, on the hope that the spread might not be so bad anyway. Many of the countries which may acquire nuclear weapons, they say, will be quite responsible, especially when equipped with so awesome a capacity for destruction.

Analysis of what it would be like to live in a crowd of countries with nuclear weapons leaves very little doubt that the spread would introduce new and very threatening dangers to the world. Even neglecting fall-out, thermal effects, and the effects of prompt radiation, the blast from a half dozen low kiloton weapons distributed on Tel Aviv, Jerusalem, Haifa, and Beersheba might kill some 800,000 persons, a catastrophe which would dwarf the 2,000 fatalities experienced in the October War, and which would hardly compensate for the million civilians who might be killed by three low-yield bombs on Cairo and one on Alexandria. The threats to initiate such disasters are not likely to be very useful as substitutes for either territory or non-nuclear force in the Middle East. The rather tired arguments about perfect safety through vastly increased mutual vulnerability have not worn very well in the case of a hypothetical world of two countries and of rationally calculating men. Universal vulnerability in a world of many nuclear powers including perhaps Khadafi is even less attractive. However, while it is very likely that there will be some further spread, how much and how rapidly is not a matter of fate, but a subject for policy. So is the management of the additional spread which does take place.

That the rate and extent of spread are not immutable is shown by the fact that past and recent plans in various countries to install power reactors and chemical separation plants have altered. In fact they change all the time, responding to economic pressures as well as to domestic and international politics. The plans which were made in 1975 and which were drawn on to construct the charts (Figures 1 and 2) displaying the " overhang " of countries with enough separable plutonium for a few bombs, and the " overhang " of countries planning to separate at least

Figure 1

The Overhang of Countries with Enough Separable Plutonium for Primitive or Small Military Forces

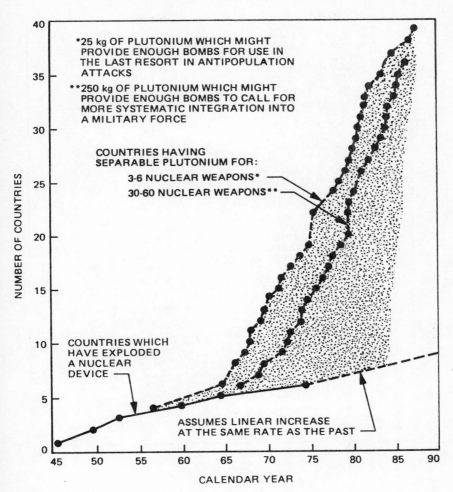

that much plutonium, were deflated somewhat to correct for overoptimism; the plans have already been changed. Some reactors have been deferred. At least one chemical separation plant—in Korea—has been cancelled. We can affect such plans by deliberately changing the economic and the political and military incentives entering into nuclear policies.

A fatalism which holds that nothing can be done today may be an unconscious cover for a desire to do nothing, to continue as before. This would be intelligible in light of the large vested interests, as much psychological as financial, in the movement towards the use of plutonium

Figure 2

Countries Planning to have Plants for Separating Plutonium or Enriching Uranium in Sufficient Quantities for Several Bombs

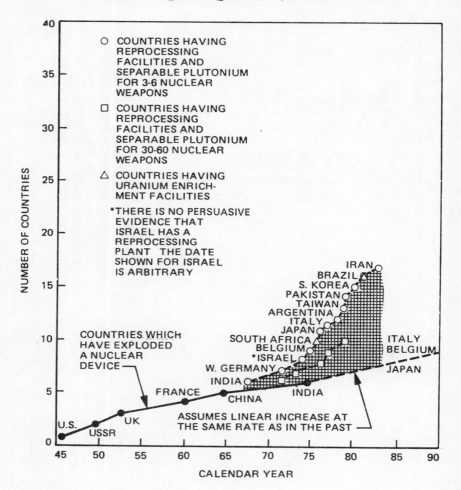

fuel. But a refusal to pause to reconsider the commitment to plutonium is likely to be short-sighted even from a parochial economic view.

The Possibility and the Need to Choose

Slowing and limiting the spread of nuclear weapons will mean shaping political desires and this, in turn, will depend on the conduct of relationships within alliances, as well as on a sober and clear-headed view of the economics and politics of the exports of nuclear materials. There is room for a choice of policies.

In the civilian use of nuclear energy, there are possible choices among distinct forms of nuclear electrical power. We can, for example, choose between nuclear power with and without the recycling of plutonium. In the military field, there are alternatives in the policy of alliance and in the development and use of non-nuclear technologies which can displace nuclear ones. And if we take " arms control " in the broadest sense to include, besides formal agreements, any actions in the joint interest of potential adversaries, there are important alternatives here too.

A few of these central points deserve expansion.

The shrinking critical time to make explosives: Past alarms have been mainly false. There have been several: the first was given immediately after Hiroshima; the next major one at the end of the 1950s. A typical prophecy of 1960, for example, forecast the addition of 12 new countries to the list of those which would make and explode nuclear devices in the following six years. By comparison with these early alarms the actual increase in the number of countries testing nuclear weapons has been very slow. Three additional countries tested at intervals of eight, four and 10 years in the 22 years following the British nuclear explosion.

The error in these early predictions of rapid spread gives some ground for doubting the apocalyptic prophecies which swarm in this field. It is clearly a matter of political will and not merely of technical and industrial competence. Nonetheless there is cause for alarm today in the changes which are gathering beneath the surface. Even if the number of countries which have tested nuclear explosives continues to develop slowly, the civilian nuclear programmes now under way make it certain that many new countries will have travelled a long distance down the path to a capacity to produce nuclear weapons. In many cases the remaining distance will be short enough to mean that even a rather small impulse might carry a government the rest of the way, and the movement of one government might provide the impetus for others. This in turn would mean a new and dangerous instability.

In fact, the fundamental overlap of the paths to nuclear explosives and to civilian uses of nuclear energy has been recognised since the mid-1940s. We have almost from the start said that the military and civilian atoms were substantially identical, yet paradoxically, that we wanted to stop the one and to promote the other.

This paradox was present in the discussions leading up to the Truman-Atlee-King declaration of November 1945. The most valiant effort to reconcile these opposing purposes was contained in the Acheson-Lilienthal proposal of 1946.

The Acheson-Lilienthal plan tried to resolve the dilemma by proposing to " denature " plutonium, *i.e.*, to make it ineffective as an explosive. This was to be accomplished by leaving the fuel in the reactors long enough so that the fissile isotope, plutonium-239, generated in the uranium fuel rods would in turn generate higher isotopes of plutonium, and in particular, plutonium-240, which was known to have several drawbacks from the standpoint of the art of weapons design of the time. The discussion of this proposal was necessarily muted and confined by the requirements of secrecy and by the current state of the art. The initial report was predicated on the belief that denaturing would impose the formidable barrier of isotope separation for the plutonium and thus, given the elaborate mechanism of international control called for in the Acheson-Lilienthal proposal, assure a warning of two to three years. This hope was almost immediately modified, but it has generated a long and inconsistent trail of statements which still have their effect in encouraging the belief that plutonium left in the reactor long enough to become contaminated with 20 to 30 per cent. of the higher isotopes plutonium-240 or plutonium-242 would be unusable, or, at any rate, extremely ineffective when used in a nuclear explosive. Since reactors operated " normally " to produce electrical power were expected for economic reasons to achieve a maximum " burn-up " of the fuel by leaving the fuel rods in the reactor long enough to contaminate the rods, a kind of denaturing was hoped for as a result of standard procedures in the operation of power reactors. However, this hope turned out to be a slender reed.

An explosive which is truly nuclear may be made from such plutonium from power reactors; it will release 1,000 times the energy per pound of ordinary high explosives. This is now clear from authoritative, publicly available documents. Therefore, a need for isotopic separation of plutonium-239 does not form a barrier to making an explosive from the plutonium characteristically produced as a by-product of the generation of electricity.

For governments accumulating the spent fuel, the barrier to obtaining a high enough concentration of fissile plutonium will be the need to separate the plutonium chemically. This is a less formidable obstacle than isotopic separation, the plant for which costs billions of dollars, using present techniques, and which would need many years for construction. The critical time to make an explosive from spent reactor fuel is less than the two or three years originally anticipated. Nonetheless, chemical separation is a substantial barrier and perhaps the most important one remaining. It might take a year to construct a chemical reprocessing plant for the purpose, and once the plant is completed, at least two months to produce plutonium dioxide from the hot irradiated fuel rods. Depending, however, on what forms of the nuclear fuel cycle become general, and where and under whose control the different phases of the cycle occur, governments may start with material considerably closer to the plutonium metal than the irradiated spent fuel. They might, in fact, without violating any existing rules, start with separated plutonium dioxide or plutonium nitrate. That might be perhaps a week or less away from the metal. The rules, in fact, do not, in general, preclude stocking the metal itself; and as Dr. Theodore Taylor has said, the oxide powder may be used directly.[18]

Figure 3 presents three possible nuclear fuel cycles for the light water reactors (LWRs), involving three alternative dispositions of plutonium. Among other things, it simplifies by not distinguishing permanent from temporary storage. The first disposition involves recycling uranium and the temporary or permanent storage of plutonium. The second involves no recycling of either uranium or plutonium and the storage of spent fuel.

[18] Willrich, Mason and Taylor, T. B., *Nuclear Theft: Risks and Safeguards* (Cambridge, Mass.: Ballinger, 1974).

Figure 3

Alternatives for the Disposition of Plutonium, 1990

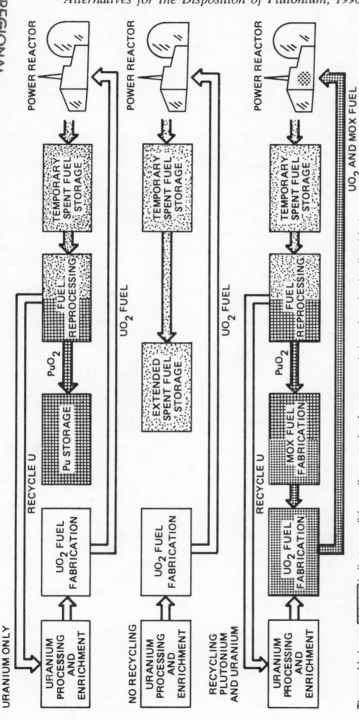

The speckled areas ⬚ indicate condition yellow: stocks of unseparated plutonium, *i.e.,* irradiated fuel.

The dotted areas ⬚ indicate condition orange: stocks of plutonium in unirradiated fresh mixed oxide fuel.

The cross-hatched areas ⬚ indicate condition red: stocks of separated plutonium.

Plutonium in these three conditions comes successively closer to availability for use in nuclear explosives.

SOURCE: This has been adapted from six alternatives considered in U.S. Atomic Energy Commission, *General Environmental Statement on Mixed Oxides (GESMO)* (Washington, D.C.: U.S. Government Printing Office 1974), Vol. I, pp. 5–50.

The third involves the recycling of both uranium and plutonium; it also involves, besides chemical reprocessing, the fabrication of " mixed oxide " (MOX) fuel.

The first is said, by the *General Environmental Statement on Mixed Oxides (GESMO)*,[19] to be the basic current practice, but in fact present practice is much closer to the second alternative, with no recycling whatsoever. The third alternative has been contemplated since the early 1950s in the United States and has served as a model in plans formed in many other countries before the soaring costs of reprocessing called it into question. Even before this increase in cost, the more optimistic estimates current at the time *GESMO* was written showed no substantial savings in costs per kilowatt hour to be obtained by recycling. Nor did they show any decisive effects in conservation. Plutonium recycling was expected to delay a shortage of uranium by perhaps a year and a half at the rate of demand anticipated for the 1990s. Since then the benefits in conservation have been estimated to be even smaller and the economic benefits appear negative.

The alternative fuel cycles have very different implications for shortening the critical time to make a nuclear explosive. In some of the fuel cycles, plutonium will exist in different forms which will vary in their degree of closeness to being available for use in an explosive. Figure 3 distinguishes three states at which plutonium might be found at different points of these cycles. The first state is in spent fuel. We have called this " condition yellow " to indicate that although obtaining the plutonium in spent fuel represents a considerable stride along the road to nuclear weapons, compared to the situation of the countries which at present possess weapons and which started from scratch, substantial further effort is required. Reprocessing of spent fuel involves remote manipulation of extremely toxic, radioactive substances, facilities with six or seven feet of shielding, lead glass windows, etc., and the handling of substantial quantities of spent fuel in order to produce small quantities of plutonium. Chemical separation of plutonium for weapons, however, can be simpler than the separation of plutonium as an economical substitute for uranium in electrical power reactors. The spent fuel can be found of course in all of the three fuel cycles.

At the other extreme is the plutonium which would be stored at the " back " end of reprocessing plants and at the " front " end of plants fabricating plutonium or " mixed oxide " fuel. Such plutonium in the form of plutonium dioxide or plutonium nitrate could be converted to plutonium metal using generally known methods and without remote handling equipment or extensive shielding and the like, but only a glove box. It should take no more than a week. We, therefore, speak of " condition red " for the points in the various alternative nuclear fuel cycles at which plutonium dioxide (PuO_2) or plutonium nitrate ($Pu[NO_3]_4$) might be found. Such stages do not occur at all in the light water reactor cycle without recycling of either uranium or plutonium. They do in the other two cases, and especially where the plutonium is recycled.

19 U.S. Atomic Energy Commission, *op. cit.*

Plutonium would also be found, if it is recycled, in fresh unirradiated fuel rods at the " front " end of the reactor. Extracting plutonium from such mixed oxide fuel would be substantially easier than taking it out of the irradiated spent uranium fuel. Plutonium is considerably more concentrated in the mixed oxide fuel rods (4·5 per cent. compared to 0·7 per cent.). Unlike the irradiated fuel, it is not highly radioactive and would require no " hot cells " with heavy shielding, remote manipulation, etc., and no removal of fission products. We designate the points at which fresh mixed oxide fuel might be found as " condition orange ".

We can measure the advance towards the ability to manufacture nuclear explosives implicit in recent civilian nuclear electrical power programmes by showing first the number of countries, including the countries

TABLE I

Plutonium Available from Reloads of Mixed Plutonium and Uranium Oxide (MOX) Fuel in the Early 1990s [a]

	Kilograms of Plutonium [b]	Equivalent Number of Bombs [c]
Austria	400	46
Belgium	2,800	325
Brazil	500	58
West Germany	11,700	1,357
India [d]	360	42
Iran	3,200	371
Italy	2,100	244
Japan	9,000	1,044
South Korea	900	104
Mexico	800	93
Netherlands	400	46
Philippines	1,000	116
Spain	5,600	650
Sweden	4,800	557
Switzerland	3,200	371
Taiwan	3,200	371
Yugoslavia	500	58
Egypt	700	81

a Using only indigenously produced plutonium and assuming that one reload is always kept at each reactor. Any country not having facilities for MOX fuel fabrication could justify this practice although this may not be standard practice in all countries on this list. Even if this is not the case, a single MOX reload would probably contain 350–900 kg of plutonium (equivalent to 40 to 104 bombs). In countries which do have mixed oxide fuel fabrication facilities, there would be still larger amounts of plutonium available in process.

b Assuming 580 kg per 1,000 MWe boiling water reactor reload and 770 kg per 1,000 MWe Pressurised Water Reactor reload, linear scaling for other reactor sizes. See U.S. Atomic Energy Commission, *General Environmental Statement Mixed Oxide Fuel (GESMO)*, (Washington, D.C.: U.S. Government Printing Office, August 1974), Vol. III, p. IV C–65.

c 8·62 kg plutonium per bomb assuming 5 kg fissile plutonium per bomb and assuming MOX plutonium is 58 per cent. fissile plutonium.

d The figures for India are the result of direct calculation.

possessing weapons at present, which will have enough separable but possibly unseparated plutonium for a few bombs between now and 1985— " condition yellow ". Then the number of countries which have planned to have a capacity to separate that much plutonium by 1985—" condition red ". And third, the quantities of plutonium which will be available in fresh " re-loads " of unirradiated fuel (the intermediate " condition orange ") to a large number of countries, if plutonium recycling should become general, and even if these countries do not themselves separate plutonium or manufacture plutonium fuel rods. (The results of these three sets of calculations are contained respectively in Figures 1 and 2, and in Table I.)

The first thing to be said about the numbers in these charts is that they are very large ones. The chemical separation of plutonium and the enrichment of uranium are " civilian activities " which have long been regarded as " normal ", if not yet operational, parts of the nuclear electrical fuel cycle. They may sometimes and in some places be discouraged by various *ad hoc* national policies, but they have not been subject to a clear-cut international or universal national prohibition by the countries which supply nuclear material. The problem of inhibiting or reducing the size of this burgeoning capacity is not merely a matter of an improved watch, to see that a clearly agreed prohibited line is not crossed. Among other things, it would involve defining and moving such a clearly understood and agreed boundary to prohibit activities which cannot provide adequate warning. And for whatever dangerous activities remain on the permissible side of the agreed boundary, we need to elaborate consistent national and international policies to discourage them and encourage other safer alternatives.

The second thing to be said is that this large growth is not inevitable. It presumes the carrying through of plans, negotiations, and constructions not yet committed, and of varying degrees of firmness; many have had setbacks. The growth, moreover, is open to influence, a subject for the elaboration of the policy of supplying as well as recipient governments. Some of the major limitations in the persuasiveness of American influence on the policies of various importing and exporting countries stem from the ambivalence, confusion, and arbitrariness of American policy on access to fissile material. Some aspects of the policy for the United States must be developed in combinations appropriate to particular countries, but without subverting the main point. The numbers in Figures 1 and 2 and Table I are not unconditional forecasts, but indications of what may happen if conditions are not altered.

For our present purposes, however, the gist of Figures 1 and 2 and Table I is that, under the present rules of the game, any of a very large number of countries may take these further long strides towards the production of nuclear weapons in the next 10 years or so without violating the rules—at least no rigorously formulated and agreed rules.

Governments may be able to appropriate stocks of highly enriched uranium or plutonium acquired for other purposes in order to develop weapons. Those without the national means to produce such stocks may have acquired them legitimately for their research facilities or as fresh fuel for their commercial reactors. A single reload for an HTGR, for example, might contain 250–500 kilograms of highly enriched, unirradiated uranium. If plutonium recycling becomes widespread, a single reload for an LWR might contain 350–900 kilograms of easily separable, unirradiated plutonium. Governments may also accumulate plutonium for the initial loadings of future commercial breeders. They do not have to steal these stocks or even divert them in order to be very close to being able to make nuclear explosives. Many bilateral agreements give title to the stocks to the recipient government. Some bilateral agreements do not explicitly exclude the use of these stocks for making " peaceful " nuclear explosives. And of course even for those countries which have ratified the Non-Proliferation Treaty, the exclusion of " peaceful " explosives stipulated in Article V would not survive the three-month notice of withdrawal permitted under Article XI.

The paths of approach towards a capacity to produce weapons considered here, it should be noted, do not break any precise, generally agreed rules. They are additional to paths which exploit the weakness of sanctions against infringements on the Non-Proliferation Treaty or bilateral rules; they are also additional to paths open to those governments which have not ratified the Non-Proliferation Treaty. Extending the Non-Proliferation Treaty to more countries or increasing the efficiency of " safeguards " or physical security measures would not, therefore, block these paths. The recent wave of interest in measures against " diversion ", while useful in itself, distracts attention from the steady spread of productive capacities within the rules.

Some of the stocks of fissile material could always be diverted within the permitted limits of error of material unaccounted for by any inspection system. In the future when these stocks are very large, even a small percentage of such diversion would yield significant amounts. This tends therefore to be the focus of most attention. Yet it may be much less important than the possibility of accumulating the whole of a significant stockpile of fissile material legitimately, without diversion, and using it later for explosives.

Our present study has distinguished for convenience four kinds of nuclear explosive capacity. The first is the sort of capacity which has been much in the public eye in the last year or two, in consequence primarily of the efforts of Dr. Theodore Taylor to make clear its dangers.[20] It would consist of the manufacture of a crude device derived from stolen fissile material, perhaps not using plutonium metal, but plutonium oxide powder, yielding as little as 10 or 100 tons of energy, and designed for terrorist use by some non-governmental group, or possibly even a single individual. It might use poorly separated material and be dangerous not merely to explode, but to store and handle.

---

[20] See, for example, the " Statement of Dr. Theodore B. Taylor, Chairman of the Board, International Research and Technology Corporation ", U.S. House of Representatives, Subcommittee on International Organizations and Movements, and Subcommittees on the Near East and South Asia of the Committee on Foreign Affairs, *U.S. Foreign Policy and the Export of Nuclear Technology to the Middle East* (Washington, D.C.: U.S. Government Printing Office, June 1974), p. 15; Willrich, M. and Taylor, T. B., *op. cit., passim*; and Wohlstetter, Roberta " Terror on a Grand Scale ", *Survival*, XVIII, 3 (May–June 1976), pp. 98–106.

The second capacity would rely on a few explosives in the kiloton range. They might be used by governments as a desperate threat against populations in the last resort. The third capacity we have taken arbitrarily as consisting of perhaps 50 such devices, enough to call for plans to incorporate them into a military force. The fourth would be much more sophisticated. It is the kind which an industrial power like Japan might contemplate, if it made the decision to become a military nuclear power in the 1980s or 1990s. .

This study focuses especially on the second and third sorts of capacity and the conditions for obtaining them. Such a capacity for desperate attacks on cities imposes no stringent requirements for delivery. (The requirements are much more stringent for a middle-sized power to acquire a serious and responsible force in the 1980s.) We do not however mean to imply that the primitive capacity to produce bombs for use against civilians will actually realise the hopes some governments might place in it. It is likely to be extremely inflexible, vulnerable, and available only for suicidal use. Nonetheless some governments might take this route. And while capabilities of the second and third sort are more elementary than the sophisticated capabilities sought by a middle power, they are not therefore less dangerous. They can be used to coerce a power that does not have nuclear weapons, and if both sides in a conflict of two regional adversaries were to possess such vulnerable forces, the situation would encourage pre-emption, and would thus be extremely unstable.

Relating Policy to " Legitimately" Shrinking Critical Time

The problem of preventing or at least slowing and limiting the further spread of nuclear weapons is clearly a complicated one. An adequate programme must deal with highly enriched uranium as well as plutonium and with research reactors and critical experiments as well as power reactors, and it should aim to improve the bilateral and international inspection systems. Moreover, such a programme must concern itself not only with restrictions on these sensitive materials, facilities and services, but also with providing safer substitutes for them at minimal economic sacrifice and on a non-discriminatory basis. Finally, it must deal with the structure of alliances and guarantees that may keep incentives low for countries to acquire nuclear weapons in their own defence. None-theless, restrictions on commerce in sensitive materials, facilities and services are a necessary though not sufficient condition for any serious programme to slow and limit the spread of nuclear weapons.

While it is not the purpose of this study to make detailed recommen-dations for policy, our analysis of trends in the spread of civilian nuclear technology does help to define some neglected problems which policy must address. It serves to show the inadequacy of many policies proposed at present.

The most fundamental problem treated by our analysis is that, for an increasing number of states not possessing nuclear weapons, the critical time required to make a nuclear explosive has been diminishing and will continue to diminish without any necessary violation of clear, agreed rules—without any " diversion " needed—and therefore without any prospect of being curbed by safeguards which have been elaborated for the purpose of verifying whether the mutually agreed rules have or have not been broken. This definition of the problem permits us at the very least to observe how much of our efforts—useful though they are for other purposes—deal squarely with this basic trouble.

We can illustrate the need for addressing that basic trouble more precisely by commenting on several current proposals for dealing with the problem of the spread of nuclear weapons. Some of the proposals at present most intensely advocated are rather fluid. For example, the form and functions of the multinational nuclear centres, recently suggested by the United States Government, have been changing rapidly. This continuing evolution makes careful attention to the relation between policy and purpose all the more necessary. Some variants of these proposals may be useful. Some are plainly irrelevant. Some might have the perverse effect all too familiar in this field.

Improvements in safeguards: The IAEA inspection system is designed to see that agreements under IAEA inspection are not violated, that materials are not " diverted " and that the limits of error of material unaccounted for are kept small. Whether or not such a system will provide early warning depends on the nature of the agreement, on what is excluded and what is permitted. If agreements are formulated so loosely as to make it perfectly legal to accumulate stocks of plutonium in a form days or hours from insertion into a nuclear explosive, no search for violation of the agreement, no matter how diligent and " tight ", will provide timely warning. If we wish to make IAEA inspection arrangements serve the purpose of warning, it is imperative to define the agreements so that a close approach to the manufacture of weapons is a violation. To say this does not weaken our support of the IAEA. It is a condition for making it effective.

There are many suggestions for improving the IAEA's system of safeguards against diversion by governments. These include an increase in the budget for inspection, *e.g.*, employing more inspectors; the improvement of the technology of " real-time, on line " inspection; and the use of more subtle system-analytical methods or game-theoretical methods and the like for detecting diversion in significant amounts. But the dangerous activities we have described shorten the time needed to make explosives without resorting to " diversion ". Hence these improvements in the amount of inspection or its statistical or technological sophistication are irrelevant for the purpose of preventing the reduction of the " critical time ". The line drawn between the " safe " activities which are permitted under agreement and the dangerous and prohibited activities needs to be redrawn and clearly defined to make safeguards useful. If, for example, the storage of separated plutonium, the manufacture of plutonium dioxide fuel rods, and the use of recycled plutonium in fresh fuel rods are banned, an increase in the budget or the efficiency of methods for safeguarding of fuel might be feasible and useful. Improved inspection might be even more effective if,

in general, countries which do not have nuclear weapons would agree
not to store spent fuel under national control. This might be accomplished, for example, by leasing rather than selling to them the spent
uranium fuel for electrical power production or research, or by the
storage at a fair price of their spent fuel in weapon states or in international centres under strict control. Since the management and disposition of spent fuel is a serious technical problem, such practices might
serve a double function.

Almost everyone is in favour of improving safeguards, even those who
resist all the restraints on technology which might make safeguards
effective. The latter sometimes object to constraints on the ground that
the spread of nuclear explosive material is inevitable. If this were true,
there would be little point in safeguards, improved or otherwise, except
as a façade for the steady spread of military nuclear technology.

" Collaborative transnational policing " by quasi-public or private groups:
This is not so much a policy put forward by any government as it is a
recurring theme advanced by natural scientists and technologists who think
of themselves as serving interests which transcend purely national ones.
This old theme has recently been revived with the suggestion that the
growth in the number of countries increasingly close to the threshold of
the manufacture of bombs will increase the opportunities for " transnational sanctions " by industrialists, and by scientists and technologists.

The suggestion for amateur surveillance raises doubts first about its vir-
tues compared to the more formal and official varieties of surveillance
and, second, as to its relevance for the problem. Transnational loyalties
have a larger scope for operation in some countries than in others. National
loyalties and national control of dissidence vary greatly from country to
country. Transnational actions blocking important governmental plans seem
rather less likely in Taiwan, South Korea, Yugoslavia, and Rumania, for
example, than, say, in Japan or Switzerland. And if any amateurs tried,
they did not succeed in curbing India's progress towards the manufacture
of nuclear explosives even before the recent suppression of dissent in India.
Most to the point here, however, is the fact that surveillance of any sort,
amateur or professional, is not enough. It does not address the problem
posed by a country's legitimate movement along the path towards
nuclear explosives. To curb such movements, one needs to redefine what
is legitimate. Only then are sanctions, formal or informal, conceivable.

Physical security against terrorists: This has been perhaps the principal
focus of public interest in recent years. The spectacular possibilities of
nuclear terror by radical groups have been conjured up by the growth in
stocks of strategic fissile material simultaneously with the actions of terrorists, especially in the Middle East and Latin America, and by the
internal disorder in the United States in the 1960s. It is only natural that
the mass media of communications, the United States Congress in its hearings, and inevitably as a result, the regulators of the Nuclear Regulatory
Commission and others concerned with civilian nuclear industry, pay a
good deal of attention to this possibility. Concern with this contingency

has been intensified by the efforts of a number of able physicists, in particular Dr. Theodore B. Taylor.[21] However, even if the possibility of theft by terrorists were completely eliminated, this would not exclude the possibility that a government might itself divert materials. Still less would it prevent a government from steadily accumulating strategic fissile material within the existing rules. Though measures of physical security may cause private industry to grumble, one of their advantages is that governments themselves are not hostile to them. Governments have a clear self-interest in keeping nuclear explosives out of the hands of conspiratorial factions and can therefore agree on a common domestic policy. A good many problems in the international field are much more recalcitrant: specifically the problem of reconciling the spread of civilian nuclear power with the restriction of access to fissile material by governments.

However, measures of physical security do not deal with the problem of the shortening critical time for governments to acquire nuclear explosives: that can happen without the breaking of any prevailing rules. An exaggerated concern with terrorism tends to distract attention from the problem of the spread to more countries of the capacity to manufacture nuclear weapons. Nonetheless, some useful measures have been suggested to increase the difficulties faced by a small terrorist band operating in hostile countries. For example, one might greatly dilute the special nuclear material or " spike " it with radioactive elements or " co-process " the spent uranium, that is, separate the plutonium and the uranium from radioactive by-products but not from each other. But these measures are not likely to present a serious obstacle to a well-equipped national laboratory. The confusion of the problem of terror with the problem of proliferation has therefore been unfortunate.

On the other hand, some measures taken to ban or discourage the shrinking critical time for governments might make much more manageable their problem of physical security against theft and terrorism: so, for example, the choice of a nuclear fuel cycle which does not involve recycling plutonium.

Agreements among nuclear exporters: The government of the United States has been trying to persuade other supplying governments to restrict their exports to countries which have ratified the Non-Proliferation Treaty, or which have at least accepted safeguards on all their nuclear activities. But as long as they are under " safeguards ", the Treaty does not prohibit technological devices or processes for separating plutonium, or enriching uranium; or stocking separated plutonium or highly enriched uranium; or making plutonium fuel rods or stocking fresh plutonium fuel rods made elsewhere. An inspector of safeguards might observe these permitted activities and only confirm that a given country by undertaking them had shortened the time to produce nuclear explosives. Similarly, a country which has not ratified the Treaty but has agreed to safeguards on all its nuclear activities has at most agreed to conduct its dangerous activities under safeguards.

[21] Willrich, M. and Taylor, T. B., *op. cit.*, *passim.*

There are important areas for potential agreement among exporters; they might, for example, agree to refrain from competition in the subsidy of nuclear exports to less developed countries. Export subsidies are a dubious form of international economic competition. However, the current practice of financing nuclear exports seems peculiarly irrational from the standpoint of the exporting country, as well as the rest of the world. It involves a transfer or gift of resources from the exporting country, which, compared with alternative investments, may retard the economic development of the receiving country, and, more importantly, is likely to have the particularly unpleasant external effect of bringing it closer to the manufacture of nuclear weapons. It should be possible for supplying countries to reach agreement not to subsidise the spread of weapons which might be used to destroy allies or friends, or even themselves.

Other possible agreements might be reached among the countries supplying nuclear equipment. Some agreements might be aimed directly at the problem presented by technological devices or processes which would so shorten the time necessary to make nuclear explosives as to make timely warning and response unfeasible. If importing countries can, without violation, enrich uranium in order to concentrate enough uranium-235 for nuclear explosives, or if they can separate plutonium in forms close to being usable in bombs, exporters should be clear that trilateral arrangements among themselves, the importers, and the IAEA, for even the most rigorous inspection, will not fulfil the original and essential intention of safeguards. They will provide no safety. Nor does it reduce the seriousness of the problem to say that the highly enriched uranium has a possible " civilian " use in research reactors or high temperature gas cooled reactors; and that the separated plutonium might be used in a critical experiment in preparation for the long-awaited coming of breeder reactors; and that the importer has given his pledge that the materials will be used only for such peaceful ends. Safeguards would be unnecessary if everyone could be taken at his word forever. Genuine safeguards which provide timely warning will not be possible unless there is enough time left after a violation and before the assembly of bombs for the detection of signals and taking action on them. For that purpose, supplying governments have grounds for restricting the sale of equipment, materials, and technology in order to prevent the shortening of the time to acquire nuclear explosives.

Neither the United States nor any other exporting country can claim always to have kept clearly and continuously in mind the essential need to restrict the spread of such nuclear activities as reprocessing, which could defeat the purpose of timely warning. In spite of the statements of certain journalists, none of these governments can justifiably claim to have been much holier than the others. Nor would it be fruitful to do so. Much more important is the fact that other exporting governments, like that of the United States, have both an interest in the net gains from trade resulting from an efficient, unsubsidised nuclear export industry and a vital interest in curbing the wide spread of nuclear bombs. For some of the sensitive technological devices and processes, the economic benefits which might be forgone are plainly minor. The gains from the exportation of reprocessing equipment, for example, are likely to be modest at best.

Reprocessing appears to be uneconomic even in very large separation plants, and much more so in the small plants required in poor countries. But even if, in disregard of economic considerations, reprocessing were to become universal, the demand for separation facilities and equipment would be small in relation to the demand for reactors—perhaps 2 per cent. of that market. Exporters might of course try to gain some special advantage in the competition to sell reactors by tying the sale of reactors to an offer to supply separation plants to importers eager to gain control of readily fissionable material. That, however, is precisely the sort of suicidal competition which must be avoided.

Assured supplies of slightly enriched uranium compared with internationally controlled supplies of mixed oxide fuel: It should be possible to assure slightly enriched fuel on a non-discriminatory basis to meet the demands of countries which observe new and improved conventions against proliferation. It is part of President Carter's programme to do so. Some of the reasons making this feasible are that the demand for nuclear electrical power, and therefore the demand for nuclear fuel, have been greatly exaggerated and have now been scaled down drastically. The outlook for a supply at a reasonable price is good for the rest of the century, and since fresh, slightly enriched uranium fuel is not an explosive material, it should be possible to arrange for substantial working stocks under effective national control. The programme is also made feasible by the fact that fuel costs are relatively small in proportion to investment costs, and, since the fuel is not bulky, storage costs are not excessive.

According to press accounts, one of the alternatives considered by the London Suppliers Group for reducing dependence on slightly enriched uranium would permit use of fresh plutonium fuel under close international control. At the recent hearings at Whitehaven on a proposed expansion of the Windscale separation plant, representatives of British Nuclear Fuels Limited (BNFL) suggested this as an alternative for governments which are concerned to achieve " energy independence ". BNFL also recommended that new plutonium fuel be placed under strict international storage and control and released only according to international criteria. This proposal would make these countries more rather than less dependent on outside sources for an uninterrupted fuel supply, and their reactor operations would be much more liable to shutdowns than with the slightly enriched uranium fuel it would be feasible and safe to supply. Presumably, BNFL's proposal would mean keeping strategic quantities of plutonium out of the hands of governments which do not have nuclear weapons. If such arrangements were practicable at all, keeping the amount of plutonium under national control to less than a bomb's worth or a few bombs' worth would allow these countries almost no working stocks of MOX or separated plutonium under their own control. With only one MOX reload as a working stock for each reactor, and assuming they do not fabricate their own MOX fuel, in the 1990s Japan and the Federal Republic of Germany would each have more than 1,000 bombs' worth of plutonium quickly accessible and even Spain would have 650 bombs' worth. (That is, on their plans up to recently. If they fabricated their own MOX fuel they would have even more plutonium, in forms still more directly used in nuclear weapons.) But less than one thousandth or one 650th of a country's annual reload requirement could hardly be called a working stock.

The American experience with India offers strong evidence that even supplies of slightly enriched uranium fuel which would guarantee operation of the Tarapur reactor for over two years have been deemed by the Indian government to be below emergency levels, dictating resupply by air and other speedy action. Moreover, the debate in the 1950s on the draft of the IAEA Statute focused on similar although less drastic proposals for the deposit of fissionable materials with the IAEA. Even then it was made clear that to give such powers to the IAEA was unacceptable to governments like India's, as threatening their economic life and their independence. It seems extremely unlikely that governments trying to secure a little more " energy independence " by the use of plutonium fuel rather than natural or slightly enriched uranium would accept a new international institution which deprived them of any significant national control of such plutonium, and which thereby made them more rather than less dependent on outside power for continuity of supply.

Fresh low enriched uranium stocks under national control are more likely to be susceptible to limitations which satisfy both the user's desire for adequate working stocks and the international community's desire to keep stocks of highly concentrated fissionable material out of the hands of states without nuclear weapons. It is also true that international control and close, even continuous, inspection of spent uranium fuel would intrude much less into the essential operation of power or research reactors, yet might serve an important function in providing early warning of diversion.

Plutonium separation plants in a few weapon states as a substitute for nationally owned plants: Dr. David Owen and others have suggested that it would be better to have a few states with nuclear weapons, such as the United Kingdom and France, provide separation services for countries without nuclear weapons than for the latter to have separation plants of their own. But it is hard to see how this would help matters at all. What makes a reprocessing plant dangerous from the standpoint of spreading nuclear weapons is its product, plutonium. If a country can obtain the product without the trouble of investing in the facility, the suggested arrangement would appear merely to simplify the task of acquiring plutonium for bombs.

It is worth noting then that a country need not have a national reprocessing plant in order to obtain plutonium in " condition red ", that is, in the form of plutonium nitrate or plutonium oxide powder or even plutonium metal, all of which are left open as possibilities in the contract contemplated between the Japanese and BNFL.

Multinational nuclear centres: A few years ago policy on anti-proliferation was moving in the direction of multinational and international nuclear centres. These were proposed originally for the conduct of chemical reprocessing or for isotope enrichment, both of which are dangerous activities. There is an apparent advantage in a collective management which would mean mutual monitoring. Such arrangements appeared to be useful, provided they were clearly and effectively designed to replace and exclude national centres for separation and enrichment. However, the proposals for multinational centres which were put forward until recently would not clearly forbid members of the consortium from construction of their own plants for reprocessing or enrichment on their own, perhaps on a smaller but nonetheless militarily significant scale. If they do not forbid nationally controlled plants from enriching uranium or separating plutonium and if they do not also prevent the accumulation of national stocks of easily used fissile materials, multinational nuclear centres could not deal with the basic problem of diminishing critical time. Nor is it clear that they would have the indirect effect of reducing incentives to acquire plants under national control. Plutonium recycling in the multinational centres appears to be uneconomical, although it is less uneconomical than it would be in national plants. But the current interest of governments in having their own national separation facilities does not derive from calculations of economic benefit.

In fact, in so far as reprocessing is made to appear to be of crucial importance in the nuclear fuel cycle—and in so far as much larger flows of highly enriched uranium come to seem important for the propagation of high temperature, gas cooled reactors in the future—sovereign states, especially the less developed ones, are likely to object to the surrender of their sovereign rights to undertake these activities on their own. Some less developed countries have argued that the prospect of multinational nuclear centres requires them to undertake a national enterprise for separating plutonium in order to train and prepare their own nationals for an active role in the multinational enterprise. In this way, the multinational centres may perversely give some appearance of legitimacy to national efforts on a smaller scale. Multinational nuclear centres would in fact provide some of the technological knowledge and skill needed for the national efforts. However, the crucial difficulty with a multinational reprocessing centre has been indicated in our comment on proposals to expand reprocessing services of the weapon states: it would legitimise commerce in separated plutonium and supply bomb material directly to states which do not at present have nuclear weapons.

In some proposals, multinational and international nuclear centres would be limited essentially to the storage of spent fuel. That would be compatible with proscribing plutonium recycling; combined with such a proscription, it would address the basic problem of preventing the acquisition by more and more governments of stocks of readily fissionable material.

The concentration in multinational nuclear centres rather than individual governments of plants for isotopic as distinct from chemical separation would have the same effect, as long as such plants produced only low enriched uranium suitable for reactor fuel. While the chemical separation of plutonium in multinational nuclear centres and its shipment to member countries would defeat the purpose of assuring timely warning, the multinational production of low enriched uranium for the reactors of member countries would not present the same disadvantages. Unlike separated plutonium, low enriched uranium is not readily fissionable, and is therefore not usable as an explosive. Centrifuge, or jet nozzle, or laser isotope separation might in the future greatly decrease capital costs and so make feasible and economic the operation of enrichment plants in many more countries. The ability to separate isotopes of uranium—and therefore to produce highly enriched uranium for bombs—may then threaten to become widespread. But the ability to design, develop, and manufacture centrifuges or laser separation equipment is not likely to be widespread. In that case it would help if the suppliers of technology for enrichment confined their exports to multinational nuclear centres or states possessing nuclear weapons, and if states not possessing such weapons got their low enriched uranium for reactor fuel from multinational nuclear centres or their traditional sources among the powers possessing nuclear weapons.

If not carefully defined in purpose, however, multinational nuclear centres could aggravate the problem as easily as not.

Stopping the "arms race" between super-powers: Resolutions proposing to reduce United States and Soviet strategic forces have been advanced recently in the United States Senate and House of Representatives as measures to reduce the likelihood of the proliferation of nuclear weapons. But whatever its other advantages, the reduction of the nuclear forces of the " super-powers " and of their expenditures on nuclear weapons, and a cessation by the super-powers of nuclear tests, etc., plainly do not have any direct relevance to the proscription of those activities of the states at present without nuclear weapons which, without breaking the rules, lessen the critical time needed to obtain a capacity to produce nuclear explosives. It is nonetheless sometimes suggested that they might affect incentives for the states without nuclear weapons to obtain them or to lessen the time between the decision to construct a nuclear weapon and the actual date when they will be able to do so. For the states without nuclear weapons, a vague undertaking by the states possessing nuclear weapons in the Non-Proliferation Treaty to negotiate in good faith to stop the " nuclear arms race " and disarm offers an ultimate excuse for undertaking nuclear armament on their own. Furthermore, this notion appeals to the sense of guilt familiar at least in the United States and the United Kingdom. However, there are no grounds for thinking that South Korea, or Taiwan, or India, whose potential interest in nuclear weapons is not directly tied to their use against either of the two super-powers, would have their interest reduced at all by even a very extensive reduction in nuclear armament by the super-powers.

II

PRIMER ON TECHNOLOGY

What it Takes to Build a Nuclear Weapon

After the political decision is made, there are three requirements which must be met in order to build any kind of nuclear device: a design for the device; a reasonable level of skill in working with high explosives, nuclear materials, fuses, etc.; and a quantity of fissile material.

Historically, some persons in the United States have had a mystical attitude towards "the secret of the atom bomb", as though there were some password or secret handclasp which would secure instant admission to the club of nuclear powers. The Manhattan Project scientists seemed to underestimate the difficulty of designing the bomb and had as their slogan, "there is no secret to the atom bomb". More recently, nuclear scientists, such as Dr. Theodore Taylor in *The Curve of Binding Energy*,[1] have emphasised the ease of construction. The basic principles of bomb design are fairly well known. A few years ago a high school prankster in Florida sent the police a letter claiming that he had fabricated a nuclear device and demanded payment not to detonate it. He enclosed a sketch of his (non-existent) device. The police checked the sketch with federal authorities, and it was realistic enough for them to feel constrained to take the threat seriously. Recently an enterprising television producer hired an MIT student to design a bomb for a special television report on proliferation. Dr. Jan Prawitz, of the Swedish Ministry of Defence, was asked to evaluate the design. His remarks were edited on the programme to make it appear that the design was a good one.

Such incidents have led many to conclude that building nuclear weapons is an extremely simple matter. This conclusion should be treated with caution. For example, in contrast to the edited remarks of Dr. Prawitz on the television programme, his full evaluation of the student's design was as follows: (1) The design of the bomb was primitive. (2) Several essential features were omitted. (3) The student might not survive the fabrication of the bomb. (4) Should he survive, the probability that the bomb would go off upon "pressing the button" would be less than 50 per cent. (5) Should a nuclear reaction start, the explosion yield would be small. The probability that the device would yield as much as ·01–1 kt would be very small. (6) If a group of qualified scientists or a professional defector were granted more time and better means, including the possibility of undertaking experiments, the bomb could be much better.[2]

Dr. Prawitz's last point needs emphasis in connection with the spread of nuclear weapons to governments as distinct from subnational groups. A well-equipped national laboratory of the kind now established in small or developing countries with scientists and engineers trained in the United States, the United Kingdom and other advanced states would find it a much easier task, given the fissile material.

[1] McPhee, John (Westminster, Md.: Ballantine Books, 1975).
[2] Letter to Mrs. Roberta Wohlstetter, 16 July, 1975.

There may be no single secret of the atom bomb, but there is a whole collection of ideas, techniques and devices, which are not public knowledge and which contribute to the ability to build with safety a reliable bomb with a reasonable yield. These ideas and techniques shade off into the area of technological skill rather than fundamental design.

The second requirement for building a nuclear device is a certain level of technological skill. The non-fissile parts of the bomb—the high explosive lenses, the fuses, etc.—require a reasonably high order of workmanship if the device is to work efficiently, but nothing beyond standards which are already available in the civilian market. The fissile materials require very careful handling because of their radioactive character. If plutonium is used, the workers will have to be protected from a highly poisonous substance. The methods for coping with plutonium are fairly well known, however, and a country with any kind of nuclear power industry should be able to build up a team of workers able to work safely and effectively with this dangerous substance. Countries with nuclear weapons, for example, have already found the means to handle large quantities of plutonium in a reliably safe manner.

The final requirement, for fissile material, is probably the most difficult one to meet. The exact amount of fissile material required to make an explosive nuclear device depends on the quality of the design, the desired yield, and on the purity of the fissile material employed. Approximately five kilograms of plutonium should be enough for a first nuclear device. More uranium-235 would be required, perhaps 15 kilograms, for an implosion device.

As everyone knows, fissile material is not easy to obtain. Only a very small fraction of natural uranium (about 0.7 per cent.) consists of the fissionable isotope uranium-235. Separating one isotope from another of the same element is a difficult and expensive physical process. Plutonium does not occur in nature, but is produced as a by-product in many nuclear reactors. The separation of plutonium from the other elements found in the used fuel rods of a reactor is a chemical rather than a physical process, and although it is not easy, it is very much easier than separating one isotope from another. The methods of acquiring fissile material deserve discussion in somewhat greater detail, since they appear to present the most difficult obstacle to development of a capability to produce a nuclear weapon.

Acquiring Fissile Material by Gift, Theft, or Purchase

One way, perhaps the simplest, to acquire fissile material is by means of gift, theft, or purchase. There are a variety of reasons why a country possessing fissile material might be willing to give some of it to one which does not possess it. It might be given under the supposition that it was to be used for research only, in order to help establish a technological case for the ultimate construction of a nuclear power industry in the recipient country. Or it could be given for political purposes, in order to stabilise a potential situation of conflict or to enhance the prestige of an ally. Or it could be simply purchased, since a country which really wanted nuclear weapons might be willing to pay for fissile material at a price sufficiently high to attract an offer of sale from one which had a nuclear power industry, but little interest in acquiring nuclear weapons for itself. Inhibiting international trade in fissile materials is an interesting challenge for future makers of policy.

Theft has ordinarily been considered more as a threat from terrorists than a threat from sovereign governments. It is not too difficult to protect nuclear fuel when it is actually in the reactor. It is, after all, in a highly radioactive environment and is protected by all the security measures taken to protect the power plant itself. Radioactive waste is also not a very attractive place from which to steal fissile material. The reason for this is that great care is taken to ensure that radioactive waste will not " go critical " under any conceivable contingencies, and therefore the concentration of fissile material in such waste is always very low. The most vulnerable link in the chain is probably while nuclear material is in transit. Whether or not the nuclear reactor fuel—prior to being " burned "—is attractive as a target for theft depends on the particular type of reactor for which it is destined : light water reactor and heavy water reactor fuel is not attractive, since it must undergo further enrichment, while high temperature gas reactor fuel—which may never become particularly common—only requires a chemical separation. After burning but before reprocessing, it is once more vulnerable and valuable. In the United States a considerable effort is being made to improve the security of fissile material. Standards of security in other countries, however, tend to be much lower. The International Atomic Energy Agency, for example, imposes no requirement of any kind on physical security against theft, although it does issue strong recommendations.

Acquiring Fissile Material by Isotope Separation

If a country is unable to procure fissile materials by gift, purchase, or theft, or if it rejects such means in favour of a more independent source of supply, one alternative which it might consider is isotope separation of uranium. This process requires, of course, a supply of uranium. It is estimated that uranium constitutes about four parts per million of the earth's crust, so it is not particularly rare. Whether a given piece of rock should be considered " uranium ore " depends, of course, on the price one is willing to pay. Table I shows how many tonnes of uranium can be extracted from reasonably assured reserves in various countries at a price of $26 per kilogram. The second column shows estimated (but not assured) additional reserves if the price moves from $26 to $39 per kilogram. As more intensive exploration is carried out in the less developed countries, we can expect that additional reserves will be discovered. So it is perfectly evident that uranium is not particularly rare in and of itself, although the ore of highly concentrated uranium is unevenly distributed. The major problem comes in separating out the 0·7 per cent. of natural uranium which is fissionable uranium-235. There are several ways of doing this, such as gaseous diffusion, gas centrifuge, the Becker nozzle, and laser enrichment.

Gaseous diffusion has been, historically, the most used technique. It is based upon the fact that lighter molecules in a gas move faster than heavier

TABLE I

Uranium Reserves [1]

Country	$26/kgU Assured (Tonnes)	$26/kgU Estimated (Tonnes)	$26–39/kgU (Tonnes)	Total (Tonnes)
1. United States	300,000	500,000	1,200,000	2,000,000
2. Canada	190,000	190,000	340,000	720,000
3. Sweden	–	–	330,000	330,000
4. South Africa	202,000	8,000	88,000	298,000
5. Australia	150,000	75,000	50,000	275,000
6. Greenland (Denmark)	8,500	–	250,000	258,000
7. France	37,000	23,000	45,000	108,000
8. Niger	40,000	20,000	20,000	80,000
9. Brazil	–	–	54,000	54,000
10. Colombia	–	–	50,000	50,000
11. Argentina	13,000	8,000	18,000	39,000
12. Israel [2]	–	33,000	–	33,000
13. Portugal	7,000	6,000	10,000	23,000
14. Gabon	15,000		5,000	20,000
15. Spain	8,500	–	10,000	18,000
16. Angola	–	–	13,000	13,000
17. Central African Republic	6,000	6,000	–	12,000
18. Italy [3]	–	10,000	–	10,000
19. Japan [4]	–	8,000	–	8,000
20. Zaire	1,500	1,500	–	3,000

SOURCES: (1) Unless otherwise noted, from Poole, L. G., " World Uranium Resources ", *Nuclear Engineering International*, XX, 225 (February 1975); expressed in tonnes of U_3O_8 (yellow cake). (2) *Nuclear Engineering International*, XVIII, 204 (May 1973), p. 391. Contained in Phosphate Rock. Ketzmel, Z., " Uranium Sources Production and Demand in Israel ", United Nations Conference on Peaceful Uses of Atomic Energy, Geneva, 1971, A/CONF 49/P/013. (3) Bullio, Pietro, " Italian Nuclear Power Industry ", *Nuclear Engineering International*, XVIII, 209 (October 1973), p. 784. (4) Kiyonari, Susumu, " Activities of the Power Reactor and Nuclear Development Corporation ", *Nuclear Engineering International*, XVIII, 206 (July 1973).

molecules, and thus strike the container walls more often. If one of the walls has holes large enough to let individual molecules through, but not large enough to permit bulk passage of the gas, more of the lighter molecules than of the heavier molecules will pass through the barrier. By this device it is possible to increase slightly the concentration of lighter molecules in the gas. The effectiveness of a separation technique is indicated by the " separation factor ", which is the ratio of uranium-235 to uranium-238 atoms after one application divided by the ratio of uranium-235 to uranium-238 atoms before the application. The theoretical separation factor for gaseous diffusion using uranium hexafluoride gas is 1·00429; in other words, it would require a minimum of 2,300 passes to convert uranium hexafluoride from 0·5 per cent. uranium-235 to 99 per cent. uranium-235. Each stage in gaseous diffusion requires a compressor to keep the gas on the input side of the barrier at a higher pressure than on the output—

enriched—side; heating elements are also required at each stage—uranium hexafluoride is solid at room temperature. The net result is that gaseous diffusion plants consume huge amounts of electrical power. The three gaseous diffusion plants in the United States—Oak Ridge, Tennessee; Portsmouth, Ohio; and Paducah, Kentucky—require 6,000 MWe of electricity when operating at full capacity. That is approximately the power consumption of the entire state of Minnesota. The capital cost of a gaseous diffusion plant is also very large. There are only two such plants now operating in non-communist countries aside from the United States: at Capenhurst in the United Kingdom and at Pierrelatte in France. A large internationally financed plant is being built by Eurodif in France which will cost 15 billion francs, not including the cost of the four 900 MWe nuclear power plants which will be built to supply it with electricity. Gaseous diffusion plants are necessarily very large and, therefore, expensive. Because of this, building a gaseous diffusion separation plant will not be an attractive way for smaller countries to acquire fissile materials.

A somewhat less developed technique for isotope separation involves the use of a gas centrifuge. If uranium hexafluoride gas is put in a rapidly rotating cylinder, the heavier molecules will tend to become more highly concentrated near the outer radius. The separation factor which can be achieved in a single stage depends on the geometry of the cylinder, its speed of rotation, and the temperature. However, separation factors of 1·1 to 1·2 have been reported. This means that relatively few stages of centrifuge processing are required to reach high levels of separation. The rate of flow per centrifuge is low, however, so that large numbers of centrifuges must be operated in parallel in order to achieve reasonable levels of production. The technology of gas centrifuges requires the ability to produce strong, light-weight cylinders which can resist the powerful centrifugal forces generated, and also the ability to produce high-quality low-friction bearings. There are three major centres of gas centrifuge technology in the non-communist world: the United States; the three countries of the Urenco-Centec agreement—Britain, the Federal Republic of Germany, and the Netherlands; and Japan. The Urenco-Centec group has two pilot plants in operation, each about four times as large as would be required by a modest programme producing five weapons a year. A third pilot plant is now under construction. The Power Reactor and Nuclear Fuel Development Corporation (PNC) of Japan is not quite so far along. PNC started operating a 180 centrifuge plant in the autumn of 1974, and a 250 centrifuge plant is expected to be running soon. It plans to start a 10,000 centrifuge pilot plant sometime in 1978. The electricity consumption of a centrifuge plant is much less than that of a gaseous diffusion plant, but its capital costs are approximately the same. It is estimated that the centrifuge separation will be comparable in cost to gaseous diffusion by the mid-1980s, but it must be recognised that the high-speed centrifuges represent a high level of industrial technology in themselves.

The Becker nozzle uses the same principle—centrifugal force—as the gas centrifuge. It has indeed sometimes been called a "fixed centrifuge". The idea is that uranium hexafluoride gas, mixed with hydrogen, is blown at high velocity around a curved track. The heavier molecules tend to go to the outside, and the flow is separated into an inner flow (slightly enriched in uranium-235) and an outer flow (slightly depleted) by a knife-edge at the end of the track. Although a pilot plant which is believed to use this process is now operating in South Africa, and a group in Germany has been promoting the construction of a pilot plant there, it will probably be many years before any substantial quantities of uranium-235 are produced by this process. It is hoped that the cost of separation may be cut by half or more by this new technique, since it does not require expensive rotating machinery or sophisticated permeable barriers, although it appears that the electrical power consumption of the technique may be at least as great as that of the gaseous diffusion method.

The final technique is laser enrichment. Successful laser enrichment for light elements was announced by both the Soviet Union and the United States in April 1975. A proposed production technique involves using a finely tuned high power CO_2 laser differentially to excite molecules of different atomic weights; this differential excitation could then be used to separate the uranium-235 from the uranium-238. The method may eventually make separation much cheaper, but it is not yet clear how many years and how many dollars will be required before the method emerges from the laboratory and goes into the pilot plant, let alone full-scale production.

To summarise, isotope enrichment is still an expensive high-technology capital-intensive process. Although there are a fair number of enrichment plants in operation—and more under construction—outside the United States (Table II), we shall see below that the chemical reprocessing of reactor fuel to separate fissile material offers a much more promising road for poorer countries than the construction of costly enrichment facilities.

Acquiring Fissile Material by Reprocessing Spent Fuel

The fuel of any reactor has to be periodically removed and replaced by fresh fuel. The " used " fuel is then allowed to " cool ", *i.e.*, a period of about 100 days has to pass so as to allow some of the more radioactive nuclides to decay; this simplifies both handling and subsequent chemical processing. After the fuel rods have " cooled " and been transported to a reprocessing apparatus, they are subjected to " head-end " treatment, which is required to reduce the many different kinds of fuel elements to a single substance ready for chemical processing. The " head-end " processes may include the disassembly of a fuel bundle, the removal of fuel-rod cladding by mechanical means or dejacketing by chemical or electrochemical methods, dissolution of fuel but not of the cladding, or complete dissolution of the fuel element. Then the chemical separation process itself may be begun. There are several different techniques to perform a chemical separation: all have as one of their goals the separation of the plutonium and uranium from all the other materials. Modern methods can separate out over 99 per cent. of the uranium and plutonium available. The fissile material may be used either to make more reactor fuel or to make nuclear weapons.

TABLE II

Some World Enrichment Facilities Outside the United States

Country	Type	Start Up	90% Enriched U/yr (kg) [a]
France			
Pierrelatte [1]	Gaseous Diffusion	1958	1,586–1,982
	Gaseous Diffusion	1981	37,668
Germany (FRG)			
G/K [2]	Nozzle	1967	7·9
Steag [2]	Nozzle	1974	–
Urenco [3]	Centrifuge	1974	99
Japan [4]			
PNC	Centrifuge	August 1974	$\approx 3\cdot5$
PNC	Centrifuge	Early 1975	$\approx 5\cdot0$
PNC	Centrifuge	1978	≈ 200
Netherlands [3]			
	Centrifuge	Early 1974	99
	Centrifuge	1977?	793
South Africa [5]			
Prototype	Nozzle?	5 April, 1975	
	Nozzle?	1984	19,825
United Kingdom			
Capenhurst [6]	Gaseous Diffusion	1953	1,586
Urenco [3]	Centrifuge	Early 1974	99
	Centrifuge	1977?	793

a Based on a 0·2 per cent. tails assay.

SOURCES: (1) Dreissigacker, H. L., " Expected Development for Enriching Uranium in Europe ", presented at the DFVLR Colloquium, Porz-Wahn, Germany, 14 May, 1970. Capenhurst Translation 249, issued by the U.S. Atomic Energy Commission as CONF-700557-3. (2) Gillette, Robert, " Nozzle Enrichment for Sale ", *Science*, LXXXVIII, 4191 (30 May, 1975), p. 912. (3) Smith, David, " What Price Commercial Enrichment? " *Nuclear Engineering International*, XIX, 218 (July 1974), p. 572. (4) " Japan—Centrifuge, Good Progress ", *Nuclear Engineering International*, XX, 225 (February 1975), p. 81. (5) " South Africa—Release More Details on Enrichment Process ", *Nuclear News*, XVIII, 8 (June 1975), p. 75. (6) Abagian, Vincent V. and Fishman, Alan M., " Supplying Enriched Uranium ", *Physics Today* (August 1973), p. 23.

The process of chemical separation of plutonium and uranium is complex, but is simple compared to isotope enrichment. Facilities for the reprocessing of nuclear fuel exist in Argentina, Belgium, France, Germany, India, Italy, Spain, and the United Kingdom. A small pilot plant exists in Taiwan. Such facilities will soon exist in Japan, and they are being discussed for Brazil and South Korea. Two or three of these processing plants reprocess only materials test reactor fuel, and will produce very little plutonium; but several of the plants will have the capacity to produce hundreds or even thousands of kilograms of plutonium annually. Where will they find fuel for these reprocessing plants? The amount of plutonium produced by a power reactor depends on a number of factors. It depends on the type of reactor, on the power or size of the reactor, and on the way the reactor is operated. The following production rates appear to be reasonable averages, based on a 75 per cent. load factor (Table III).

TABLE III

Average Annual Production of Separable Fissile Plutonium from Power Reactors

Reactor Type	Annual Fissile Plutonium Production (Kg per MWe per Year)
Boiling Water Reactor	·23
Pressurised Water Reactor	·237
Pressurised Heavy Water Reactor	·29
Gas Cooled Reactor	·25

SOURCE: U.S. Atomic Energy Commission, *Nuclear Power Growth*, 1974–2000, WASH 1139–74 (Washington, D.C.: U.S. Government Printing Office, February 1974), p. 24, Table XII.

These figures assume normal operation of the reactor for the production of power. The quality of plutonium produced could be increased by shorter periods of exposure and more fuel reprocessing per unit time, but there would be significant economic costs for such procedures. We can use these factors to estimate the amount of plutonium which will exist in the spent fuel elements for each country at any time over the next 15 years, on the basis of the country's existing and planned power reactors. Experience has shown that there is usually some delay in the dates of starting for power reactors, and that this delay is greater in the less developed countries. Therefore, we assume that a reactor which is planned to begin operation in the year Y will actually begin operation in the year Y + 2 if it is in Latin America or Asia, and in the year Y + 1 if it is in Europe or Canada. Reactors already in operation, of course, were credited with their actual starting dates. The estimates given may be too high, however, because of the rather optimistic assumption of a 75 per cent. load factor. High load factors are desirable in power reactors, and a 75 per cent. factor is often assumed in planning, but in practice power reactors frequently fall short of it either because of breakdowns, faulty maintenance, or lack of demand for electricity—India's nuclear power reactors, for example, have operated at less than 50 per cent. of capacity. So one should not place too much confidence in the exact figures listed in Table IV. But even if these figures are high by a factor of two, or more, the basic conclusion remains the same: the growing nuclear power industry is going to be producing a very large amount of plutonium in a very large number of countries. Although the plutonium will require chemical reprocessing to separate it from other materials in the spent reactor fuel elements, this chemical reprocessing is a much simpler operation than isotope separation.

The plutonium from a reactor which has been operated in such a way as to maximise the output of electrical power will be less than ideal

TABLE IV

*Projected Accumulated Separable Fissile Plutonium
from Power Reactors*

	Separable Fissile Plutonium Accumulated (kg)				
Country	1970	1975	1980	1985	1990
Argentina	0	0	350	1,228	2,898
Austria	0	0	0	1,324	3,059
Belgium	0	0	1,272	4,953	9,858
Brazil	0	0	85	866	3,591
Bulgaria	0	0	711	2,291	3,871
Canada	90	1,665	4,938	14,283	27,075
Czechoslovakia	0	75	358	1,984	3,639
Denmark	0	0	0	324	1,134
Egypt	0	0	0	216	756
Finland	0	0	294	1,997	3,877
German Democratic Republic	51	170	1,188	2,838	4,488
German Federal Republic	277	1,657	7,621	21,638	38,927
Hungary	0	0	0	632	1,422
India	18	402	1,032	2,424	4,204
Iran	0	0	0	1,242	5,022
Israel	0	0	0	108	648
Italy	485	935	1,941	6,099	10,774
Japan	74	1,460	10,126	23,671	35,915
Korea (Republic of)	0	0	281	1,951	6,081
Mexico	0	0	56	1,111	2,221
Netherlands	8	153	628	1,103	1,578
Pakistan	0	61	201	605	1,405
Philippines	0	0	0	226	1,356
Portugal	0	0	0	216	1,296
Romania	0	0	0	316	711
South Africa	0	0	0	513	2,220
Spain	12	588	2,613	4,562	16,836
Sweden	0	212	3,169	9,244	15,629
Switzerland	5	622	1,671	6,366	11,371
Taiwan	0	0	206	1,961	6,241
Yugoslavia	0	0	0	555	1,110

SOURCES: " World List of Nuclear Power Plants ", *Nuclear News Buyers Guide 1975*, XVIII, 3 (February 1975), pp. 45–56; International Atomic Energy Agency, *Power and Research Reactors in Member States 1974 Edition* (Vienna: IAEA, 1974); United States Atomic Energy Commission, *Nuclear Power Growth*, 1974–2000, WASH 1139–74 (Washington, D.C.: U.S. Government Printing Office, February 1974), p. 24, Table 12.

from the standpoint of the manufacture of weapons. The reason for this is that only two isotopes of plutonium, plutonium-239 and plutonium-241, are fissionable by thermal neutrons. The presence of substantial quantities of the isotopes plutonium-240 and plutonium-242 introduces three problems for the designer of weapons:

(1) Increase in critical mass. The exact critical mass required for a nuclear explosion when one is dealing with a mixture of isotopes requires some moderately complex calculations. But a single bench mark case may give the reader a sense of the order of magnitude of the problem: one would have to dilute fissionable plutonium with over 50 per cent. non-fissionable isotopes in order to increase the required critical mass by a factor of three. An increase in the required critical mass will induce a roughly proportionate increase in the mass of the total device.

(2) Increase in radioactivity. Since plutonium-240 and plutonium-242 fission spontaneously, plutonium containing these isotopes will be more radioactive than "weapons-grade" plutonium. This means that the fabrication of weapons will be somewhat more difficult, and also that a weapon, once produced, will be more detectible and thus not as suitable for clandestine delivery as a weapon made from purer material. These considerations may be important inhibitions to potential nuclear terrorists, but they are not particularly relevant to most national governments.

(3) Decrease in the reproducibility and efficiency of yield. The spontaneous fissioning of plutonium-240 and plutonium-242 increases the possibility of pre-initiation, which will have the effect of making the yield achieved by a given device somewhat unpredictable and less efficient. The quantitative analysis of this phenomenon is classified, but one can say that the uncertainty in yield introduced by reasonable quantities of dilutents is small compared to the difference between nuclear and non-nuclear weapons.

There are at least two ways in which a country could choose to get around these problems. The concentration of militarily undesirable isotopes increases as the burning of fuel increases; therefore a country could get some plutonium of high quality by accepting an uneconomically low burning on some fuel loadings. One of the major aims of the various agreements on safeguards and inspection procedures is to prevent this sort of manipulation of the fuel cycle. The other alternative is to accept the penalties associated with "normal" spent-fuel plutonium, on the grounds that a nuclear weapon which is less than optimal is better than none. Whether this latter course would be acceptable depends, of course, on the political and military motivations which drive the country in question to decide to build a nuclear bomb in the first place, but it is easy to think of circumstances in which a bomb made from "reactor grade" plutonium would be just as useful as a more refined bomb.

Categories of Competence for a First Nuclear Test

The countries of the world can be classified into categories according to the ease with which they could produce a nuclear device.

The first category consists of those states which have built and tested nuclear weapons, and which have military forces capable of the delivery of such nuclear weapons. There are today five states in this category: the United States, the Soviet Union, Great Britain, France, and China. From a political, military, and economic standpoint there are obviously tremendous differences between the countries within this category; and in particular there are great differences between the significance of their various nuclear programmes. The three " middle " nuclear powers have manifestly less strength than the two " super-powers "; it should be made clear to everyone that the mere possession of nuclear weapons does not make a country equal to the super-powers.

A subcategory of this first category consists of those states—India at present is the only one—which have detonated a nuclear device, but which claim not to have produced nuclear weapons. There is a deplorable tendency to think of " proliferation " in terms of just one weapons-test, and to forget about all the subsequent developments which are required to achieve something beyond a mere terrorist or desperate " last resort " action.

The second category consists of those countries which have full access to the fissile material required to make a weapon and a sufficiency of manpower trained in nuclear technology. The lower limit on this category may be somewhat vague, but probably most would agree that Japan, West Germany, South Africa, Belgium, Taiwan, Italy, the Netherlands, Canada and Sweden should be included. Some would argue that Canada and Sweden should not be included, because they have neither an enrichment nor a reprocessing plant within their borders. This is true, but both these countries have such a very high level of nuclear expertise that they could construct a small reprocessing plant very rapidly if they wished to. Taiwan has a " pilot " fuel reprocessing plant, which clearly demonstrates its mastery of the technology required to separate plutonium from used reactor fuel. Some persons say that Israel has some clandestine reprocessing facility and has prepared to assemble nuclear weapons very quickly if the need should ever arise.

The third category consists of those states which have at least a reactor within their borders and some base in nuclear technology. The states in this category at this time appear to be Israel, Argentina, Switzerland, Egypt, Spain, South Korea, Indonesia, East Germany, Czechoslovakia, Australia, Pakistan, Iran, Norway, Brazil, and Mexico. Some of these states do not have power reactors but only research reactors. Research reactors can be used to produce quantities of plutonium which, while not large compared to those produced by power reactors, would still be adequate to support a small nuclear weapons programme. For example, the two Norwegian research reactors generate about 3·9 kg. of plutonium annually, which is enough to build about four bombs every five years.

The fourth category consists of all other countries. One must realise that there is a continual tendency, even if they have no desire for nuclear weapons, for countries to move towards the higher categories of competence. This is the natural consequence of the expanding desire for nuclear electrical power.

How Competent is Each Category?

Now let us discuss what these different categories of competence mean if a country decides that it wants to build a nuclear weapon.

Suppose a country in the second category suddenly decides that it is in its national interest to design and test a nuclear weapon. If it had already made extensive preparations for this decision, if it had " kept open the option ", then the time required might be very short indeed, and the costs involved might have been essentially already borne. But let us suppose that no such special preparations had been made. Then there are four steps which the country would have to carry out: rework the fissile material required for the bomb; design the bomb; fabricate it; and prepare and conduct the test. " Rework " means converting the fissile material to be used into metallic form: this step is not strictly necessary, but it would make designing a bomb easier, could be accomplished within a year, and would only cost about $200,000. (All costs here are engineering estimates based on American experience. They should be considered as rough approximations.) Design of the weapon could be done, crudely, within a year at a cost of about $200,000. Ease of fabrication would depend on whether uranium or plutonium were used. Uranium is easier to work with, but plutonium is easier to acquire. It is easily within the capacity of a country in this category to work with either metal. The preparation and actual fabrication of a weapon might take a year and cost another $200,000. The power in question might wish to test the device above ground, but even if it felt constrained to do only underground testing the cost of a modest testing programme would not exceed $250,000 or six months of preparation. All four steps could be carried out in parallel rather than in series, so we must conclude that any country in the second category is within a million dollars (or less) and a year of testing a nuclear device if it makes the decision to do so.

A power in the third category which decides to design, build, and test a nuclear device faces a basic decision right at the start: whether to attempt to acquire its fissile material by reprocessing spent reactor fuel or by enriching natural uranium. The former approach is more practical. If a country decided to build a reprocessing plant just to support its weapons programme, it could do so fairly cheaply. By not recovering the uranium, and cutting radiation protection and waste management operations to the minimum required for separating plutonium, experienced persons in industry have estimated that an adequate plant could be built for around $50 million and be operating in about four years' time. It also must be remembered that estimates of capital costs are notoriously subject to error, especially in the absence of a detailed design for the plant in question. The other steps—reworking design, fabrication, and testing—would be the same as for a power in the second category; they could be done in parallel with the construction of the reprocessing plant and thus the total time consumed would be only about four years, and the cost in dollars would be only about $51 million.

If a country in the third category wanted to follow the enrichment route rather than the separation route it would undoubtedly take longer and cost more. The most sensible kind of enrichment plant for such a country to build at this time would be a gas centrifuge, which would take about five years and cost $100 million or even more. The level of technological skill required is probably outside the reach of most—but not all—countries in the third category at present.

A country in the fourth category which decided to build a nuclear weapon would probably choose to build a reactor and a simple reprocessing plant. Skill in handling nuclear materials would follow as a by-product, but many years and many millions of dollars would be consumed in the process. A reasonable " minimum estimate " appears to be about six years and about $200,000 million. It must be remembered, however, that such estimates are much less reliable than, say, estimates of the time required to go from the second category to the first. It should also be remembered that the time and resources required would be drastically reduced if fissile material was given or sold to the country in question. Although such an action may seem unlikely at the present time, it will become less unlikely as the number of countries possessing fissile material grows.

These categories should not be regarded as a ladder which is being climbed by each country of the world towards an ultimately inevitable nuclear weapon; certainly a country in the second category would require less incentive actually to build a nuclear weapon than a country in the fourth category, but the likelihood of a particular country building a nuclear weapon can only be determined by a balanced consideration of both costs and incentives. The case of Israel, a power in the fourth category, illustrates the point very well. There are rumours that Israel is just a " turn of the screwdriver " away from possessing nuclear weapons. We do not know if these rumours are true or false; the fact that they may be true shows that if the incentives are great enough even a small country with a modest nuclear industry might build nuclear weapons.

The real problem of proliferation today is not that there are numerous countries " champing at the bit " to get nuclear weapons, but rather that all the non-nuclear countries, without making any conscious decision to build nuclear weapons, are drifting upwards to higher categories of competence. This means that any transient incentive, in the ebb and flow of world politics, which inclines a country to build nuclear weapons at some point in the future will be just that much easier to act upon. The general manner in which France " went nuclear " is apt to be repeated in the future :

French interest in the possession of a nuclear weapon is readily comprehensible. What is less easily understood is the manner in which France decided to produce the bomb. For it was not a single decision, a clear-cut long-range policy rationally planned and executed, but rather a series of events and decisions—or, perhaps, lack of decisions—which led to the Sahara test in 1960. The most candid statement to the effect that France drifted toward the possession of an atomic bomb without the project ever receiving official sanction at the cabinet level, is found in the following statement which issued from a ranking member of the Quai d'Orsay and of the Atomic Energy Committee, " On the political level there had been no doctrine of French nuclear armament. In fact the manufacture of an atomic bomb, on which the work began well in advance of the decision which was not finally taken by the Government until very recently, wedged itself into our public life as a sort of

by-product of an officially peaceful effort, there existing no overview of
the problems involved, nor of the means necessary to solve them, nor of the
results to be expected. Until a very recent date we found ourselves in the
paradoxical situation of a country which already spent by virtue of . . .
accords . . . between the Commissariat and National Defense important sums
in view of a program of nuclear armament without the Government having
taken the decision to make the weapons and also without a debate . . . in
Parliament to approve such a decision ".[3]

³ Scheinman, Lawrence, *Atomic Policy in France Under the Fourth Republic* (Princeton.
N.J.: Princeton University Press, 1965), pp. 94–95.

ON KEEPING "DANGEROUS" ACTIVITIES IN CHECK

IT would be nice if there were one clear dividing line between useful, safe, nuclear activities and dangerous ones: on one side, safe, economic, civilian, peaceful activities which are to be promoted, and on the other dangerous, military activities threatening war, a net drain, even in peace-time, on the productive life of humanity and therefore to be not merely discouraged, but banned. Something like that dichotomy has been implicit again and again, from the beginning, in all the broad and in many specific statements which call for choosing civilian uses of nuclear energy instead of military uses. If the choice were that simple, there would be little difficulty in deciding to use nuclear energy for " good " rather than " evil ", or in drawing the line between the permitted and the impermissible uses, in watching for any trespass, detecting it in plenty of time, and bringing adequate sanctions to bear. But unfortunately all these good things do not come in one cluster distinct from the bad ones.

The Inadequacies of the Conventional Dichotomies of Nuclear Activities

Some civilian nuclear activities are not economic. They are economically inferior to alternative uses of the same resources; they involve large " opportunity costs " and are a net drain on economic development. Worse still, some civilian nuclear activities which are economic might nonetheless be so dangerous as to warrant a ban. Some have an almost immediate military application.

The reduction to absurdity of the dichotomy came with the proposal (self consciously called " Plowshare "), to make " peaceful nuclear explosives ". These " plowshares ", in all essentials, are swords. To push this familiar example an inessential bit further, suppose that a " plowshare " device were most economic if manufactured in aerodynamic form; suppose moreover that local and transport costs were such that it would be cheapest for each country to manufacture it locally; and suppose that " plowshare " explosives were to beat all other methods for moving earth for canals or the like and were in great demand at the price at which they could then be supplied. In that case, it would be economic—in the precise sense of representing the most productive of the alternative uses of the same resources—to make nuclear explosives for civilian use which could also be directly used without any change as bombs. They would be, in short, economic, civilian, and of immediate military use. And definitely not " safe ".

Agreements to supply nuclear materials characteristically make explicit that such aid is to further peaceful uses only. But the plain fact that any development of nuclear explosives has a military application continues to be denied, ignored or confused by India, Brazil, Pakistan, and other countries. However, " plowshare " devices are near enough to weapons to be recognised as such in the Non-Proliferation Treaty. States which do not have nuclear weapons are asked to abstain from making not only bombs, but " peaceful " nuclear explosives as well. Yet in principle the difference between " plowshare " and some other peaceful nuclear activities is merely a matter of the distance in time for converting the civilian activity to direct military use. This distance can be, and is increasingly, extremely short. There are not two atoms, one peaceful and one war-like. They are the same atom. From the start of attempts to establish international control, this was recognised to be the " heart of the problem, the close technical parallelism and interrelation of the peaceful and the military applications of atomic energy ".[1] By a curious twist, however, we have from the start discussed the promotion of the peaceful uses of the atom as if that would automatically displace the military use. That hopeful displacement was presented in its baldest form in the initial speech of President Eisenhower launching the " atoms for peace " programme as an alternative to the " atomic armaments race ". He proposed that the nuclear powers should start the programme by donating some uranium which would thereby be unavailable for weapons, and so begin a kind of nuclear disarmament. The displacement was symbolic in both quantity and quality. The transfer, by comparison with the fissile material content of the bomb stockpile, was miniscule. Its supposed contribution to " disarmament " was quietly forgotten as transfers to civilian technology began to disperse the foundations for nuclear armament on a worldwide scale.

This wishful confusion was present in the earliest declarations of policy by President Truman and Prime Ministers Attlee and King and it persists in the Non-Proliferation Treaty, and in the bilateral agreements of all supplying countries with countries importing nuclear technology. Before and after Eisenhower, clarification has been in order on the criteria of choice and the trades among various civilian uses of nuclear energy. One way or another from the very start those who had advocated promoting the peaceful uses of nuclear energy have had to face the problem of how to do it so as not at the same time to spread nuclear weapons. Evasive rhetoric on the line between safe and dangerous activities has not helped. Nor has the sentimental identification of " civilian " with " economic ", nor the rhetorical exaggeration of the immediate promise of economic uses to foster the rapid development of less developed countries. Still more specific distinctions, for example, between optimal economic alternatives and the next best use of resources, need to be made for making sensible trades. Even if it should turn out eventually to be economic to recycle plutonium in countries which now have no nuclear weapons, we have to ask how much difference recycling would make compared with alternative arrangements for nuclear electrical power which do not call for recycling, and how important indigenous reprocessing facilities would be in economic terms today. Again, some long-forgotten engineering analyses, using criteria too narrow to include such requirements as preventing access to explosive material, may have suggested the present design for high temperature gas cooled reactors (HTGRs) which use uranium fuel enriched to 93 per cent. in uranium-235; but how much would one lose in efficiency by using lower

[1] Oppenheimer, J. Robert, " Atomic Explosives ", *The Open Mind* (New York: Simon and Schuster, 1955), p. 9.

percentages of enrichment, making the HTGR fuel much less ideal for explosives? Such trades are of the essence. Yet denying access to readily fissionable material has never systematically been made part of the criterion of choice in the design of reactors or of the nuclear fuel cycle as a whole, or indeed in the development of policy on nuclear exports.

The heart of the problem of making sensible national decisions and fair international bargains entails the recognition of distinctions going far beyond the resounding simple dichotomies. Some of the problems we are running into now, in persuading less developed countries to forego development of their own capacities for chemical reprocessing, have their origin in the excessive rhetoric of the past about how essential such processes are to the nuclear fuel cycle, and how crucial nuclear energy is to their catching up with the advanced countries.

Early Troubles with the Lines between " Safe " and " Dangerous " Activities

The Acheson-Lilienthal report and other early efforts to achieve international control started with the double political commitment made by the United States with the United Kingdom and Canada on 15 November, 1945, in Washington, in the Three Nation Agreed Declaration on Nuclear Energy, to seek . . . " international arrangements to prevent the use of atomic energy for destructive purposes and to promote the use of it for the benefit of society ".[2] The problem was how to reconcile the prevention of one with the promotion of the other, given their far-reaching identity.

The Acheson-Lilienthal report—the bulk of which was written by David Lilienthal, Robert Oppenheimer, Charles Thomas, Chester Barnard, and Harry Winne—took the optimistic and somewhat paradoxical view that the closeness of " destructive " and " beneficial " uses would help control.[3] It reasoned that if control were simply negative, *i.e.*, if it consisted merely in international inspection of nationally owned dangerous atomic plants, it would never work. One could not get first-class scientists to serve as policemen, so the inspection would be poor; and national rivalries and considerations of national sovereignty would make the system unenforceable. The report proposed therefore that an international authority, the Atomic Development Authority, promote civilian nuclear energy, and that it own—and in fact have a monopoly of—all dangerous civilian activities, which would have to be distributed in a " balanced " way throughout the territories of the various countries. The balance, however, would be determined by considerations of security rather than of economics, as the late P. M. S. Blackett noted. These dangerous activities monopolised by the authority would include a very large range from mining to the operation of many power reactors, including all breeders. For example, besides owning the mines it was to own about half the nuclear electrical capacity. The other half, in national hands, would obtain its fissile material only in " denatured " form from the plants owned by the authority. Because

[2] U.S. Department of State, *A Report on the International Control of Atomic Energy*, Publication No. 2498 (Washington, D.C.: U.S. Government Printing Office, 16 March, 1946), p. 1.
[3] *Ibid., passim.*

the international authority would have a monopoly of the essentials for the conduct of civilian nuclear energy, it could recruit first-class scientists, not simply as guardians, but to do research at the frontier of the technology and to help in its transmission; and sovereign states would have an incentive to cooperate with the international authority.

Like the original Acheson-Lilienthal proposal in this one respect, the United States Atomic Energy Commission combined in its domestic programme regulatory and developmental functions for civilian nuclear energy, but experience made it clear that this was not a tenable combination. The purpose of the recent Energy Reorganization Act in the United States was to separate these functions and divide them between the Nuclear Regulatory Commission and the Energy Resource and Development Agency and to make the promotional activity more even-handed by requiring attention to non-nuclear fuels. Strong traces of the original proposal, however, survive in the present international institutions: the International Atomic Energy Agency (IAEA) " promotes " civilian nuclear energy while safeguarding materials against diversion. The inspectors offer technical advice and help; other IAEA staff-members spend the largest portion of the IAEA budget to provide information, technical assistance, and encouragement for developing nuclear electrical power—especially in the poor countries.[4] The basic statute of the IAEA assigns it the task of seeking " to accelerate and enlarge the contribution of atomic energy to peace, health, and prosperity throughout the world ". But the arrangements for control today depart a long way from the original ideal. The IAEA does not own, but simply inspects; and " dangerous " as well as " safe " activities are both governed by sovereign states. This is precisely the arrangement which the original report said was unenforceable.

However, the problem of finding a dividing line between " safe " activities and " dangerous " ones which has preoccupied the writers of bilateral and trilateral agreements on cooperation and of the Non-Proliferation Treaty, formed the crux of the difficulties in the Acheson-Lilienthal report. And, in fact, some residual traces of belief in a solution proposed there persist, namely the possibility of " denaturing " fissile material such as plutonium by adding to it an isotope which has the same chemical properties, but which spoils it as an explosive. The idea persists that plutonium can be made safe. In particular, the notion persists among high officials in governments, in international agencies, and in nuclear industry, that electrical power reactors generate plutonium-239 so contaminated with undesirable higher isotopes of plutonium that the plutonium-239 cannot be used effectively in an explosive without isotope separation. That belief underlies much of the carelessness and incoherence of the policy of the nuclear suppliers. It is worth considering in some detail, then, the origins and evolution of the idea of denaturing. In part it stemmed from the very early belief in an extreme scarcity of recoverable uranium-235—only enough in the United States for example to replace the equivalent of a few months of power consumption—and the consequent necessity to breed

---

[4] See " Nuclear Inspector or Salesman? ", *The Economist.* CCLVIII, 6914 (28 February, 1976), p. 74.

plutonium and use it as a fuel if nuclear electrical power were to have any future at all. Plutonium therefore had to be made safe.

" Denaturing " to Assure Warning of Danger

The Acheson-Lilienthal report proposed to regard power reactors putting out as much as 1,000 megawatts as safe, provided their power was developed from the fission of " denatured uranium-235 and plutonium ".[5] If the report had been correct, the operation of such a reactor would indeed have been safe, in the sense that it would take two or three years to build a plant capable of removing the unwanted isotopes. But there were hints in the report itself of some unease about what " denaturing " accomplished, and a committee of distinguished physicists, including Robert Oppenheimer himself, on 9 April, 12 days after the publication of the Acheson-Lilienthal report, made clear some important qualifications about the effectiveness of " denaturing ". First, while it was true that constructing a plant capable of separating the isotopes would take between one and three years, ". . . unless there is a reasonable assurance that such plants do not exist, it would be unwise to rely on denaturing to insure an interval of as much as a year ".[6] And it went on immediately to an even more important qualification about which we know a good deal more today: " In some cases denaturing will not completely preclude making atomic weapons, but will reduce their effectiveness by a large factor." [7]

---

[5] Though hope for it was disappointed, what was meant by " denaturing " plutonium-239 is comparatively clear. It is not so clear today what was intended by " denatured uranium-235 ". Natural uranium contains 99·3 per cent. uranium-238 and only about 0·7 per cent uranium-235, and so cannot be exploded. Uranium-238 is not readily fissionable by slow neutrons. Natural uranium, then, might conceivably be said to be denatured, so to speak, by nature. But this is rather strained and it is possible that constraints of secrecy at the time contributed to the unclarity. The passage as a whole reads:
" More marginal from the standpoint of safety, but nevertheless important, is another case of an operation which we would regard as safe. This is the development of power from the fission of denatured uranium-235 and plutonium in high power-level reactors. Such power reactors might operate in the range from 100,000 to 1,000,000 kw. If these fissionable materials are used in installations where there is no additional uranium or thorium they will not produce further fissionable material. The operation of the reactors will use up the material. If the reactors are suitably designed, a minimum of supervision should make it possible to prevent the substitution of uranium and thorium for the inert structure of the materials of the reactors. In order to convert the material invested in such reactors to atomic weapons, it would be necessary to close down the reactor; to decontaminate the fissionable material of its radioactive fission products; to separate it, in what is a fairly major technical undertaking, from its denaturant; and to establish plants for making atomic weapons. In view of the limited amount of material needed for such a power reactor, and of the spectacular character and difficulty of the steps necessary to divert it, we would regard such power reactors as safe provided there were a minimum of reasonable supervision of their design, construction, and operation. If the material from one such reactor (of a size of practical interest for power production) were diverted, it might be a matter of some two or three years before it could be used to make a small number of atomic weapons." U.S. Department of State, A Report on the International Control of Atomic Energy . . ., p. 28.
As in the case of uranium-235, fuel slightly enriched in uranium-233, which is produced from thorium, would require isotopic separation from uranium-238 before it could be used as an explosive. In this way it differs from plutonium fuel. Denatured uranium-233 also seems to have been contemplated in 1946 and is once again under active consideration as an antiproliferation measure.
[6] U.S. Senate, Committee on Government Operations, Peaceful Nuclear Export and Weapons Proliferation: A Compendium (Washington, D.C.: U.S. Government Printing Office, April 1975), p. 200.
[7] Ibid., p. 200.

This second qualification went somewhat further than the report, which had stated that the reactor material would be " . . . unusable by any methods we now know for effective atomic explosives ",[8] without removing the " denaturants ", *i.e.*, without separating and removing most of the unwanted isotopes.

Most important, that second qualification about denaturing was not itself as strong as is warranted by the facts as they are known today. Both the report in March and the press statement in April by the committee stressed that the line between " safe " and " dangerous " activities might alter with improved knowledge about the design of nuclear explosives. And there was a good deal which had not been explored even about the properties of the weapons-designs of that time. According to Robert Bacher,[9] the former head of the Bomb Physics Division at Los Alamos and a member of the committee to discuss denaturing, the discovery of the high spontaneous fission rate of the plutonium coming out of the reactor at Oak Ridge, by comparison with the plutonium which had been derived from a cyclotron, caused a crisis at Los Alamos during the summer of 1944. It led to the abandonment of " gun-type " weapons for plutonium and a great intensification of the effort on implosion. Over half the Physics Division was diverted to the study of implosion. It led also to the use of plutonium which had a very small percentage of the spontaneously fissioning higher isotopes and to a concentration on a workable design for implosion of such nearly pure material.

The work in 1946 on what was safe and what was dangerous material from the standpoint of usability in bombs was based pretty much on the state of the art in 1945, and on the state of knowledge about that state of the art. Even for the early implosion designs it seems the more cautious statement by the committee on 9 April that denaturing would not preclude, but would reduce the effectiveness of an atomic explosion " by a large factor ", was, as things developed, not nearly cautious enough. It suggested that the expected yield and destructive effect of an explosive made with plutonium from power reactors would be negligibly small. Oppenheimer, in the lecture delivered a few weeks later on 16 May, 1946, expressed the same view in more explicit form: " Even plutonium can be doctored—not without prohibitive cost if it is to be completely nonexplosive—but to be made a relatively very ineffective explosive." [10] But in actuality, according to the public authority of the former long-term director of the Los Alamos Theoretical Physics division, Dr. Carson Mark, the explosion from " essentially any grade of reactor-produced plutonium " would be truly " nuclear ",

    [8] U.S. Department of State, *A Report on the International Control of Atomic Energy . . .*, p. 26.
    [9] Interview, August 1975. See also Hewlett, R. G. and Anderson, O. E., Jr., *A History of the United States Atomic Energy Commission*, Vol. I: *The New World, 1939/1946*, (University Park, Pennsylvania: Pennsylvania State University Press, 1962), p. 251; Hawkins, David (ed.), *Manhattan District History Project Y: The Los Alamos Project*, Vol. I: *Inception until August 1945* (Los Alamos, New Mexico: Los Alamos Scientific Laboratory of the University of California, 1961): reproduced in facsimile by the United States Atomic Energy Commission, Division of Technical Information, Oak Ridge, Tennessee, 1975, pp. 81 ff.
    [10] Oppenheimer, J. R., *op. cit.*, p. 8.

that is " an explosion of at least three orders of magnitude more energy per pound than would be available from high explosives ".[11]

The yield of any explosion is not exactly determined. There is a probability distribution of yields, a distribution with a sharp peak and a narrow dispersion in the case of a standard design using materials with few impurities. For a similar bomb using power-reactor plutonium, the dispersion might be larger, low yields more probable, the probability of a failure larger, and the expected or average yield might be lower. That the probability distribution of yields would make it nonetheless a formidable nuclear weapon is evident from the observations of Dr. Mark and Dr. Theodore Taylor, the former deputy director of the Defense Atomic Support Agency, and an outstanding designer of nuclear weapons at Los Alamos from 1946 to 1956.[12]

Moreover, the reduction in destructive effect would be smaller than any reduction in the yield of energy. The area destroyed by the overpressure of the blast diminishes as the two-thirds power of the reduction in yield, and the reduction in prompt radiation—the dominant effect on human beings of a low-yield weapon—is even smaller. (If the expected yield were eight kilotons and the less probable but actual yield were " merely " one kiloton, the area destroyed by blast would be reduced not by seven eighths, but only by three fourths and the region in which persons in residential buildings would receive a lethal dose of prompt radiation would only be halved. The lethal area would still be nearly 28 million square feet or almost a square mile.)

The statement that " . . . denaturing . . . will reduce . . . effectiveness

---

[11] Feld, B. T., Greenwood, T., Rathjens, G. W. and Weinberg, S. (eds.) *Impact of New Technologies on the Arms Race*, Pugwash Monograph (Cambridge, Mass.: M.I.T. Press, 1971), pp. 137–138.

[12] McPhee, John, *The Curve of Binding Energy* (Westminster, Maryland: Ballantine Books, 1975), p. 214; Taylor, Theodore B., " Diversion by Non-Governmental Organizations ", in Willrich, Mason (ed.), *International Safeguards and Nuclear Industry* (Baltimore and London: Johns Hopkins University Press, 1973), p. 181.

Oppenheimer, immediately after the Trinity test, estimated that the Nagasaki bomb had a 12 per cent. chance of performing less than optimally, a 6 per cent. chance of yielding less than 5 kilotons and a 2 per cent. chance that it would be under 1 kiloton, but that the lowest yield under predetonation would not be much less than that: " Memorandum: Oppenheimer to Farrell and Parsons, 23 July, 1945 ", in *TS-Manhattan Engineering District Papers*, Box 14, Folder 2, Record Group 77, Modern Military Records, National Archives, Washington, D.C. More impurities, *i.e.*, a higher fraction of plutonium-240 and -242, would alter the probability distribution of yields, but would not change the fact that at its least the yield of an implosion device of this very first type ever developed would be reliably between one and 20 kilotons.

Until the autumn of 1976 public doubts about this matter may have been justified, since only qualitative information had been made easily available, and official statements had for many years been inconsistent as well as vague. For a history of such statements from 1945 to 1976 see Wohlstetter, Albert, *The Spread of Nuclear Weapons: Predictions, Premises and Policies* (Los Angeles: Pan Heuristics E-1, 1976), pp. 60 ff. However, some precise quantitative information was released in three places in the autumn of 1976: Wohlstetter, Albert, " Spreading the Bomb Without Quite Breaking the Rules ", *Foreign Policy*, 25 (Winter, 1976–77); Gilinsky, Victor, Commissioner, United States Nuclear Regulatory Commission, " Plutonium, Proliferation and Policy ", lecture delivered at the Massachusetts Institute of Technology, 1 November, 1976 (typescript); Selden, Robert W., Lawrence Livermore Labs, " Reactor Plutonium and Nuclear Explosives," paper presented to representatives of the American Nuclear Society and Atomic Industrial Forum, Washington, D.C. reported by Gillette, Robert, " Military Potential Seen in Civilian Nuclear Plants ", *The Los Angeles Times*, 26 June, 1977.

by a large factor", while vague, was unfortunately misleading. What is more, the statement referred to primitive designs, and the authors of the Acheson-Lilienthal report themselves stressed that the state of the art in the design of weapons would improve and would always require a reconsideration of the line between " safe " and " dangerous " activities : " Only a constant reexamination of what is sure to be a rapidly changing technical situation will give us added confidence that the line between what is dangerous and what is safe has been correctly drawn; it will not stay fixed." [13] Oppenheimer, having said that " . . . plutonium can be doctored . . . to be made a relatively very ineffective explosive ", added that it would also be :

. . . a difficult one to use in the present state of the art. I don't need to tell you that the art may change and that no kind of control is worth anything which doesn't make provision for such change. It's not only that it can; it probably will, in one way or another. [14]

On the matter of the likelihood of changes in the state of the art as in other aspects of the problem of safeguards, the Acheson-Lilienthal report shows some ambivalence. While the report said unequivocally that " the limit between what is safe and what is dangerous . . . will not stay fixed ", it also said that : " . . . the development of more ingenious methods in the field of atomic explosives which might make this material effectively usable is not only dubious, but is certainly not possible without a very major scientific and technical effort." [15]

Needless to say, the art of weapons design has greatly advanced in the intervening 30 years, and much information relevant to such design is now publicly available. Dr. Mark has concluded :

I would like to warn people concerned with such problems that the old notion that reactor-grade plutonium is incapable of producing nuclear explosions—or that plutonium could easily be rendered harmless by the addition of modest amounts of plutonium-240, or " denatured ", as the phrase used to go—that these notions have been dangerously exaggerated. [16]

What gives this history contemporary relevance is the persistence, in the face of such warnings, of the belief that chemically separated plutonium with a substantial percentage of the higher isotopes such as plutonium-240 and plutonium-242, along with plutonium-239, is of rather negligible value as an explosive material and therefore is not very dangerous. This residue of the early faith in denaturing lives on in recent statements that the plutonium produced in power reactors is barely conceivable as an explosive material. For example, a recent article belittles the dangers of the reprocessing plant sold to Brazil by the Germans. After indicating that the small enrichment plant would have " little use for military purposes ", it goes on to say that the final stage of the reprocessing plant is the only place " where one would find plutonium of a grade that might

[13] U.S. Department of State, *A Report on the International Control of Atomic Energy* . . . , p. 30.
[14] Oppenheimer, R. J., *op. cit.*, p. 8.
[15] U.S. Department of State, *A Report on the International Control of Atomic Energy* . . . , pp. 26, 27.
[16] Mark, Carson, in Feld, B. T., *et al.*, *op. cit.*, pp. 137–138.

conceivably be used in an explosive ", but that the existence of such plutonium separated from the spent fuel of a power reactor " is not half the problem " that it is thought to be.[17] Officials of Kraftwerk Union, the German firm which participated in the agreement with Brazil, have said much the same, as have high officials in British nuclear agencies. In a similar vein, the French ambassador to the United Nations explained that even if plutonium from a reactor sold by Framatome to South Africa were recovered by costly reprocessing it " could not be used for military purposes ".[18] Even officials of the United States Government in agencies granting loans and subsidies to countries like India, which already possess or intend to construct reprocessing plants, take comfort from the fact that:

While the plutonium produced by these reactors could be used in an inefficient and unsophisticated explosive program, it is not optimum material for explosive uses because of the high percentage content of the non-fissionable plutonium isotope plutonium-240.[19]

It is a long way from the original concept of denaturing to the notion of making do with less than " optimum material ", although today we persist in talking about the material which is somewhat less than optimal as if it were safe. That conviction continues to influence policy on nuclear exports.

The Gravity of Danger Signals, Warning Time and Response

The initial efforts at international control in the Acheson-Lilienthal report are of interest today in another respect. They show that the problem of determining the conditions under which sanctions might be applied has turned out to be in some ways unexpectedly harder than was originally

---

[17] Rippon, Simon, " Brazilian Bonanza ", *Nuclear Engineering International*, XX, 231–2 (June–July 1975), pp. 497 *et seq.* Mr. David Smith, the new editor of *Nuclear Engineering International*, continues in the same vein as Mr. Rippon. In remarking on the opposition of Senator Symington and others to the export of enrichment and reprocessing facilities to Brazil, he refers to the " misconceived notion that nuclear energy technology is in some ways related to the hazards and horrors of ' The Bomb ' . . . [as] a fallacious mental association ": *Nuclear Engineering International*, XXI, 248 (September 1976), p. 5.

[18] " Situation in South Africa ", Address by Louis de Guiringaud before the United Nations Security Council, 19 June, 1976 (New York: Service de Presse et d'Information, 1976), no. 76–95, p. 3. M. Guiringaud was made Foreign Minister in August 1976. For similar statements by officials of Framatome, see *Nucleonics Week*, XVII, 23 (3 June, 1976), pp. 1–3. See also the recent report by the Swedish government committee on radioactive waste which says, " The plutonium . . . produced in Swedish power reactors contains as much as 25–30 per cent. of plutonium-240 . . . [and] can only be utilized in weak and probably unreliable nuclear charges of highly questionable military value ": " Spent Nuclear Fuel and Radioactive Waste," *Statens offentliga utredningar 1976: 32* (Stockholm: Liberforlag. 1976), p. 43. Similar German views are suggested by the tables presented by Dr. G. Hildebrand of Kraftwerk Union in a lecture to embassy officials of many countries at Muelheim on 30 April, 1976, on " *Brennstoffkreislauf und Exportsituation* " (typescript). See also the " Statement of Submission " by British Nuclear Fuels Ltd. at the Public Local Inquiry into an Application by BNFL for Permission to Establish a Plant for Reprocessing Irradiated Oxide Fuels . . . at Windscale, June, 1977 (typescript).

[19] Letter of 18 August, 1975, to Congressman Findley from Denis M. Neill, assistant administrator for legal affairs, Agency for International Development, Department of State. For similar views of the American industry, see the Atomic Industrial Forum's Committee on Nuclear Export Policy, *U.S. Nuclear Export Policy* (Washington, D.C.: Atomic Industrial Forum, Inc., 21 July, 1976). p. 3.

foreseen; this problem is especially grave when violations or dangerous acts are less serious—or at any rate less immediately, even if ultimately, dangerous. What should we do about activities which step by step cumulatively defeat the purpose of assuring safety, when none but the last small individual step presents an unambiguously final danger? This is the characteristic way in which the dangers contained in the spread of nuclear weapons appear today. It is part of the meaning of what we have called "the shrinking critical time", or the increasing "overhang" of countries with a very short critical time for making a nuclear explosive.

Bernard Baruch's few modifications in the Acheson-Lilienthal report became part of the official United States' proposal to the United Nations Atomic Energy Commission; his key change in the report had to do with just this matter of sanctions. It was also one of the decisive points of contention with the Soviet Union.

The authors of the Acheson-Lilienthal report tended to think that the signals we would receive under their plan, when a country undertook banned dangerous activities, would in effect be signals of the beginning of an atomic war. When they talked of safeguards therefore and what we should demand of them, they said:

It must be a plan that provides unambiguous and reliable danger signals if a nation takes steps that do or may indicate the beginning of atomic warfare.[20]

The security which we see in the realisation of this plan lies in the fact that it averts the danger of the surprise use of atomic weapons. The seizure by one nation of installations necessary for making atomic weapons would be not only a clear signal of warlike intent, but it would leave other nations in a position—either alone or in concert—to take counter-actions.[21]

While the danger of an atomic attack is even more fearsome than the danger from the spread of nuclear weapons, it also makes easier the determination of how to respond to the unambiguous detection of a violation. Oppenheimer was able to say that there was a strong sanction in the Acheson-Lilienthal plan against violation, namely war.[22] The doubts which bothered Eberstadt, Baruch's deputy, and the contrasting viewpoint of Oppenheimer, Thomas, Acheson, and Herbert S. Marks, who was special assistant to Acheson, are illustrated in an account of the conference at Blair-Lee House on 17 May, 1946, where Baruch and his aides raised the subject of sanctions with the Acheson-Lilienthal group.

What do we say to those who violate? There must be sanctions, retaliation . . . Eberstadt said it was essential to state the truth about the plan: It did not eliminate the bomb. You do not eliminate the bomb merely by specifying a penalty for violations, Lilienthal replied. The public had the wrong impression, Eberstadt insisted. People thought the State Department plan was a system for controlling the bomb. Thomas asked what Eberstadt considered necessary. Outlawry and a penalty, he replied. The board's plan did have a penalty, said Oppenheimer—war. Marks pointed out that the United States would fight

[20] U.S. Department of State, *A Report on the International Control of Atomic Energy* . . . , p. 9. On the subject of danger signals, as in the case of "denaturing" plutonium, the report shows some unease and lack of clarity.

[21] *Ibid.*, p. 53.

[22] Hewlett, R. G. and Anderson, O. A., *op. cit.*, p. 565.

if Russia attacked Alaska. So under the proposed system: violations of the plan would be equivalent to violation of our boundaries.

Finally, Acheson declared there were only two ways to go further than the Board of Consultants had. One was a treaty requiring that an offender nation meet with an automatic declaration of war by all the others. He thought this meant little. The other way was world government envisioning all wars as civil wars. This meant not a " damned thing ". [23]

So long as one were thinking of extreme and immediate dangers, waging war in response to them might have been plausible with or without a formal provision in the treaty to that effect. Acheson was right that the declaration would add very little.

What if the dangerous activity is not so immediately dangerous? What if it consists simply in piling up a little separated plutonium for use in some future peaceful experiment? What if that stock of plutonium could greatly advance the development of the capacity to make nuclear weapons if and when a country decided to do so? (This need not signify any firm intention of ever making a nuclear attack.) Response then is likely to be, and has been, more uncertain as well as considerably milder than going to war.

Baruch's team, which always had a bad press among the scientists who had been members of the Manhattan Project, was concerned about the possibility of smaller violations than war and, in spite of some impression to the contrary, it did not propose nuclear attacks against violators for minor infractions. The old-fashioned and ominous term used by Baruch —" condign punishment "—has an aura of summary and extreme punishment no matter what the provocation. But it literally means exactly the opposite—only the punishment which is deserved or fitting. Baruch's suggestions were all necessarily vague, although Baruch wanted very much to make them specific. He submitted to the United Nations Atomic Energy Commission proposals that the treaty include such " enforcement provisions " as:

(1) Definition of conduct constituting violation; (2) Consequences of such violations, including the procedures to be followed in detecting, establishing, remedying or punishing such violations; (a) Administrative action by the Authority; (1) Special investigation; (2) Revocation or denial of licenses; (3) Other action; (b) Resort to judicial processes and procedures; (c) Reference of serious violation to the Security Council of the United Nations. [24]

It is easy to see why Baruch was against having the Security Council operate on these matters under the veto system. Even the administrative actions by an international authority to undertake a " special investigation " might well have run into politically determined split votes which would render it ineffective. The experience with the various international control commissions in the last 30 years, from one of which Canada recently withdrew in frustration, shows how hard it can be even to investigate an alleged violation.

[23] *Ibid.*
[24] U.S. Department of State, *The International Control of Atomic Energy: Growth of a Policy*, U.S. Memorandum, No. 1, 2 July, 1946, submitted to UNAEC Subcommittee No. 1 (Washington, D.C.: U.S. Government Printing Office, 1946), p. 151.

The ambiguity as to what is safe and what is dangerous makes for hesitation in applying any sanctions at all. The international judicial processes which might be resorted to, then as now, are notably slow and ineffective. A decision of the World Court coming several years later is likely to be split, reflecting at least the ambiguity, as well as the political differences. The withholding of licences might have been effective if the international authority had a true monopoly. But it raised the question as to who would control the authority: a sore point for the Soviet Union. It is easy to see why the Soviet Union, which felt acutely American dominance in the Security Council, was against having the Council operate on these matters without a Soviet power to veto.

In the end, the sort of activities which increase the ability of a national government to detonate a nuclear explosive may not violate any strict promises; or if they do, they may do so ambiguously and in a way which seems threatening only in a distant future. To such dangers no responses more belligerent than the withholding or withdrawal of some benefit in the form of nuclear materials and services seem at all plausible, and even these are likely to be easy to argue against.

Civilian Benefits and National Sovereignty

The prospect of economic benefits has played a major role both in spreading nuclear technology and in the hope of avoiding the spread of nuclear weapons. Suppliers have offered civilian technology as an inducement to governments to foreswear the production of bombs and to accept safeguards, and in this way they have implied that the materials and services needed for civilian nuclear energy programmes be withheld or withdrawn as a potential punishment for violations of promises to foreswear the production of nuclear explosives. However, whether as a prospect to be enjoyed or denied, these supposed civilian benefits have been exaggerated.

As for the benefits, the present value of civilian nuclear energy in particular for the economic development of the poor countries has been greatly inflated and this exaggeration has only made countries not possessing these weapons less willing to accept any restraints on their civilian programmes. In fact, it has led paradoxically to their insistence that, in return for a quite revocable promise not to make or accept nuclear explosives, they be given civilian technological resources which would carry them a long distance towards the capacity to produce such nuclear explosives.

As for the denial of benefits, the prospect of withdrawal of assistance in the development of civilian nuclear energy has become less likely and fear of the prospect has also diminished, for several reasons: because the benefits were not as large or as imminent as had been hoped; because the withholding could be made to appear as a violation of the pledge to offer assistance; because an increasing number of suppliers appeared who were competing in offers of such assistance; and finally, because within the supplying governments, there developed factions with a vested interest in continuing the supply in spite of violations.

The Power of Positive Thinking about Atomic Power

The shock at the destruction of Hiroshima and Nagasaki aroused a strong impulse towards finding an equally constructive or positive side to the release of nuclear energy. This is most often attributed to the guilt feelings of the scientists of the Manhattan Project. Robert Oppenheimer was very conscious of this impulse and cited Enrico Fermi's wry comment on the subject:

" It would be nice ", Fermi said, not quite without seriousness, " if it could cure the common cold ". He was talking of atomic energy. It was not yet three years since he had established the first fission chain reaction. Now we were meeting at Los Alamos, in the tired summer of 1945, in the week in which Hiroshima and Nagasaki were bombed. We were meeting at the request of Secretary Stimson's Interim Committee, not for the first time, and not for the last. We were being asked about the future of research and development in atomic energy, not only on the military side, but also, and even more urgently, with regard to its peaceful applications. Mr. Stimson had just issued a statement, after the bombing of Hiroshima, saying that the release of atomic energy faced us with far more than the development of a new weapon. Of course he was right. Among the peaceful applications, among the " benefits for mankind ", would be the generation of atomic power: electric power from the energy of large-scale, controlled nuclear fission.[25]

This impulse was by no means restricted to the scientists and engineers. At a meeting of the Interim Committee which considered the use of atomic explosives in Japan, Secretary of War Stimson " spoke continuously about ways to use nuclear energy for other things ' than killing people '. He wanted the members to get started immediately ' talking and questioning' about methods ' to take advantage of the changed relation of man to his universe '. He wished them to begin to figure the means by which ' the peace of the world would be helped in becoming secure ' ".[26] Moreover Secretary Stimson at that time looked forward to a drastic transformation of world politics which would bring about a supranational order hardly less utopian or more precisely conceived than the hopes for international order which were expressed by the less worldly scientists. His memorandum to the President of 25 April, 1945, had said that the " new energy could not be brought within the now obsolete schemes—nationalism, secret diplomacy, the balance of power. It required a direct and open approach, an offer for the mutual limitation and control of ' the benefits of future developments . . . ' ".[27] He recognised that the Soviet Union was a stumbling block, but hoped that future civilian benefits could be exchanged for the renunciation of the military applications of nuclear energy.

Prime Minister Clement Attlee drafted a letter to President Truman in the summer of 1945 in much the same vein. He spoke of " a new valuation of what are called national interests ", [28] and invoked " an act of faith "

[25] Oppenheimer, Robert, " The Environs of Atomic Power ", in the American Assembly (ed.), *Atoms for Power: United States Policy in Atomic Energy Development* (New York: Columbia University Press, 1957), pp. 15–16.

[26] Quoted in Morison, Elting E., *Turmoil and Tradition: A Study of the Life and Times of Henry L. Stimson* (Cambridge, Mass.: Houghton Mifflin, 1960), pp. 635–636.

[27] *Ibid.*, p. 637.

[28] Gowing, Margaret, *Independence and Deterrence: Britain and Atomic Energy 1945–1952*, Vol. I (London: Macmillan, 1974), pp. 65 and 81.

by the United States and the United Kingdom. Even so tough-minded a man as Ernest Bevin, then Foreign Secretary, early in October favoured giving the Russians the " knowledge of the atomic bomb process ",[29] and trusting their good faith in observing future international controls. Bevin's inclination did not last through October, but it is interesting that Dean Acheson, whom revisionist historians regard as a prime originator of the " cold war ", like Bevin, originally favoured making substantial disclosures of information to the Russians as part of a programme of exchange to build mutual trust, much like that advanced by the utopian scientists' movement at the end of the war.[30]

The British decided that, " while waiting for mutual trust to be established and Utopia to arrive ",[31] they would make their own atomic position as strong as possible. The British reaction to the panel's proposal for the Acheson-Lilienthal report in the following spring is suggested by Professor Gowing's summary: " In two months the panel produced a report in which they believed passionately; in this they differed from the authors of nearly all the other reports on control that had been drawn up." [32]

Many of the activities connected with the production of nuclear electrical power were in fact " dangerous " and the British government was apprehensive about the Acheson-Lilienthal report on political grounds:

To transfer development activities to an international authority would mean in practice that atomic energy plants would be built in all the principal countries with the advice and assistance of other nations. Whether the advantages of such a scheme would justify the inherent risk depended on political judgment, and certainly no such scheme could succeed except in an atmosphere of confidence. There were, however, such wide differences of opinion and outlook and policy between the great powers that there was little prospect of general acceptance of the ideal of a supranational authority whose behests would automatically override sovereignty.[33]

The British reservations were nothing compared to those of the Russians. P. M. S. Blackett, writing from a communist standpoint, advanced the view that civilian nuclear energy was likely to be of particular importance to the development of the Soviet Union and of the poor countries.[34] He thought they would not entrust so critical a part of their future economic development to an anonymous international agency, the decisions of which were not subject to veto. This was a major theme of the Soviet criticism of the Baruch plan. Blackett, who opposed a programme for producing British nuclear weapons, was of course much more extreme than the British government on the subject of international control. He hinted that the

[29] *Ibid.*, p. 67.
[30] " Forrestal Diaries " (unpublished manuscript), Office of Naval History, Washington, D.C. See Acheson, Dean, " Memorandum to President Truman on U.S. Policy Regarding Secrecy of Scientific Knowledge about the Atomic Bomb and Atomic Energy ", U.S. Department of State, *Foreign Relations of the United States*, 1945, Vol. 2 (Washington, D.C.: U.S. Government Printing Office, 1967), pp. 48–50. Interview with Paul Nitze, August 1974.
[31] Gowing, M., *op. cit.*, p. 72.
[32] *Ibid.*, p. 88.
[33] *Ibid.*, p. 89.
[34] Blackett, P. M. S., *The Military and Political Consequences of Atomic Energy* (London: Turnstile Press, 1948), p. 99.

Soviet acceptance of the Baruch plan would be interpreted as a sign of such weakness that the Americans might wage a preventive war against the Soviet Union.[35] Amusingly enough in the light of the current attack on American public utilities as an irresponsible and self-interested lobby for nuclear energy, Blackett suggested that the American utilities were behind the desire to stifle the development of civilian nuclear energy,[36] which he believed would be an essential result of the Baruch plan.

Such official and unofficial foreign reactions to the early American efforts to offer the alleged benefits of nuclear energy for civilian uses, in exchange for the surrender of national sovereignty to an international authority, called into question some of our basic initial assumptions. The optimistic and paradoxical view of the Acheson-Lilienthal report that the extensive overlapping of the destructive and beneficial uses of nuclear energy would further international control rested in part on the belief that, if the benefits were very large, sovereign states would surrender many of their sovereign rights to obtain them. But the more important and immediate these civilian benefits were made to appear, the more reluctant these states were to give up sovereign control over the civilian developments or to accept restraints on them simply to avoid supposed military dangers.

In the early period when the United States was most intensely focused on its conflicts with the Soviet Union, it tended to neglect the fact that newly sovereign states, like India, would guard even more jealously their " inalienable " sovereign rights to pursue a programme for civilian nuclear energy so hopefully conceived. Yet " atoms for peace " was in large part directed at the " power-starved ", poor countries. The Non-Proliferation Treaty was a highly ambiguous and uncertain set of compromises, embodying but not resolving the same dilemmas about national sovereignty and the problem of encouraging the production of civilian nuclear energy, while discouraging the production of nuclear weapons.

Present hopes for multinational nuclear centres may founder on the same rocks. Sovereign rights to separate plutonium or to enrich uranium are less likely to be surrendered the more they appear to be essential for civilian nuclear energy, and the more civilian nuclear energy seems to be important for future economic growth. The less developed countries in particular will not accept a permanently subservient role in any technology which is said to be critical for their development. If reprocessing is critical —or if we or they act as if it is critical—they are likely to treat multi-lateral reprocessing as a step towards national reprocessing, or even to claim that multilateral centres require national centres as a precondition. Thus, in a perverse way, all too familiar in this field, the point of the multilateral form would be frustrated.

The irony of course is that the crucial economic significance of nuclear energy, especially for the poor countries, was and is exaggerated. If we had been free of the emotions aroused by Hiroshima, we might have more easily questioned whether subsidising civilian nuclear energy was the way to stop the spread of the military technology. Since civilian and military nuclear energy programmes overlap so extensively, a more plaus-

[35] Gowing, M., *op. cit.*, p. 89.
[36] Blackett, P. M. S., *op. cit.*, p. 95.

ible course might have been to support research and development on the improvement of fossil fuels or of other alternatives such as solar or geothermal power. It is only recently, however, that this course has begun to be followed in the United States.

Even at the beginning there was a small but impressive minority who argued for something like such a course, at least in the sense of avoiding the forced feeding of the production of nuclear energy. In England Lord Cherwell,[37] and his sometime antagonist, Sir Henry Tizard,[38] as well as Sir George Thomson and the economist Sir Roy Harrod [39] and in the United States several scientists and engineers who had been associated with the Manhattan Project, and a succession of American economists,[40] have tried to put the economic prospects for civilian nuclear energy in a more sober perspective.

Nuclear Disarmament and the Problem of "Equity"

The drive to promote the use of nuclear energy for civilian purposes in order to displace the military application and to make international control more attractive faded in the late 1940s, along with the receding prospects for international control and with the intensification of the antagonism between the United States and the Soviet Union: the conviction of Alan Nunn May and Klaus Fuchs, for espionage on behalf of the Soviet Union, the communist *coup d'état* in Czechoslovakia, the Berlin blockade, the first Soviet nuclear tests, the invasion of South Korea—all showed that there was no chance of acceptance by the Soviet Union of any serious system of international control of atomic energy. Late in 1948, the Atomic Energy Commission had advanced a programme for developing an experimental breeder, a materials testing reactor, a liquid metal cooled power reactor by the General Electric Company, and a prototype of a submarine thermal reactor by Westinghouse. The reactor for submarine propulsion was brought to full power in 1953, the year when the Korean armistice was signed and the year of President Eisenhower's speech on "atoms for peace".

President Eisenhower's speech revived and inflated hopes for the worldwide benefits to be obtained from civilian nuclear energy. Though nominally a move towards nuclear disarmament, it was actually associated with the acceptance of nuclear weapons as "conventional". The same speech states that "atomic weapons have virtually achieved conventional status within our armed services" and shows a relaxed acceptance of both the permanent importance of the role of nuclear weapons and an increase in the American reliance on them. The United States had decided at about the same time to base NATO defence policies on nuclear weapons. Secretary Dulles had advanced the "massive retaliation" doctrine. The govern-

[37] Parliamentary Debate (Hansard) 5th Series, CXXXVII, House of Lords, 16 October, 1945, col. 291 (London: H.M. Stationery Office, 1945).

[38] Gowing, M., *op. cit.*, pp. 90–91.

[39] Oliphant, M. L., *et al.*, *The Atomic Age* (the Halley Stewart Lectures for 1948) (London: George Allen and Unwin, 1949), pp. 52–80.

[40] Cuthbert, D. and Squires, A. M., "The International Control of Safe Atomic Energy" *Bulletin of the Atomic Scientists*, III, 4–5 (April–May 1947), pp. 111–116 and 185. For the economists, see esp. Jacob Marschak's two articles in the *Bulletin of the Atomic Scientists*, I, 5 (February 15, 1946) and II, 5–6 (September 1, 1946).

ment looked forward to reducing the defence budget as a result of an increased reliance on nuclear firepower. (While the efforts in 1945 and 1946 at international control were based on rather utopian hopes for surrender of national sovereignty by the Soviet Union and other countries, " atoms for peace " was advanced as an instrument in the cold war competition with the Soviet Union for the allegiance of the Third World. C. D. Jackson, who had been Eisenhower's adviser on psychological warfare was a key figure in propagating " atoms for peace " and also invented the slogan " a bigger-bang-for-a-buck " to sell the new reliance on nuclear weapons as an economic measure.)

This did not mean of course any diminution in the number of international meetings with invocations on all sides to disarmament and a halt to the nuclear arms race. But the " atoms for peace " proposal which brought about the establishment of the International Atomic Energy Agency differed drastically from the Acheson-Lilienthal and Baruch plans on the crucial matter of nuclear disarmament. In 1946 the United States was the only country possessing nuclear weapons, although its stock of weapons and weapons components had been kept remarkably small—much lower than is generally realised. By 1953 there were three countries which had made nuclear weapons, and both the United States and the Soviet Union were shortly to acknowledge that this was a distinction which could not in practice be undone. Robert Oppenheimer was right in saying that unlike the Acheson-Lilienthal report, the " atoms for peace " plan had no " firm connection with atomic disarmament "; and that its bearing on the prospect of nuclear war was " allusive and sentimental " rather than " substantive and functional ". In July 1955 President Eisenhower admitted at Geneva that : " We have not as yet been able to discover any scientific or other inspection method which would make certain of the elimination of nuclear weapons. So far as we are aware no other nation has made such a discovery." [41] The Soviet Union admitted the same somewhat more obliquely; they referred to " the existence of atomic and hydrogen weapons, in respect of which the institution of international control is particularly difficult ".[42] While the United States held out the hope that further study might discover effective methods of inspection, in fact developments since have made it even plainer that there is no genuine prospect for developing a politically feasible method of inspection which would have a significant chance of uncovering nuclear weapons, the diameters of which may be measured in inches and whose length might be two or three feet, with weights of less than one hundred pounds [43] in territories of many millions of square miles. Any attempt to stop the spread of nuclear weapons must recognise that such a halt would involve a durable distinction between states possessing nuclear weapons and those without them. References to " equality " and " non-discrimination " cannot conceal this fact.

[41] U.S. Department of State, *Documents on Disarmament 1945–1959*, Vol. I : *1945–1956*, Publication No. 7008 (Washington, D.C.: U.S. Government Printing Office, 1960), p. 512.

[42] *Ibid.*, p. 465.

[43] McPhee, J., *op. cit.*, p. 107.

The Ability to Bomb Populations as an " Equaliser " of States

The formative period for the policies of "atoms for peace" in the 1950s was also a time when it was expected that the new technology of fusion weapons would result in bombs in the 25 megaton range, with areas of destruction measured in hundreds of square miles, and when the technology of ballistic missiles was expected greatly to increase errors in the delivery of bombs—up to median circles of error of 50–75 square miles in area. In fact, the chief significance of fusion technology was to make weapons of small and medium yield cheaper and lighter; the increase in error of delivery was transient. Contemporary understanding in the mid-1950s of the strategic implications of the technology of fusion and ballistic missiles renewed the initial association of nuclear bombs on Japan with the destruction of cities. The association had been challenged at the start of the decade. But by the end of the decade, some of the most influential views again regarded "population-bombing" as inevitable, and possibly even "optimal" for purposes of deterrence. In this period also, the best known rationalisations for spreading nuclear weapons, those of General Pierre Gallois, seized on the technical developments, as they appeared then, as establishing that, on the one hand, no government could depend on any other to deter a nuclear attack on itself, since the only reply would be tantamount to suicide; and that on the other hand, any country with nuclear weapons could undertake its own suicidal defence.[44] According to Gallois, all countries are equal in vulnerability in the face of the new weapons. "One can hardly see the difference in nature between the suicide that the U.S. would commit if her policy of deterrence failed and that of France if her strategy was no more fortunate." [45] General Gallois saw part of the truth: policies which threaten only a suicidal reply are no less convincing in "self-defence" than in fulfilment of a guarantee. But neither are they more convincing. This line of argument neglects the weakness of any sustained suicidal threat and the possibility of a more difficult but more persuasive response which avoids suicidal bluffs. Nonetheless, the currency of the notion that a few nuclear weapons aimed at an adversary's cities are enough to protect one's "vital interests" is a powerful impulse to proliferation. It leads naturally to the belief that the surrender of the rights to acquire so apparently useful a force calls, in the name of equity, for compensation.

Much of the discussion of discrimination in connection with the Non-Proliferation Treaty rests on the casual assumption that the nuclear weapons which would be forsworn by various states would provide those states with unambiguous safety. Hence civilian benefits are demanded in compensation—an economic gain in return for a surrender of supposed security. In fact, the acquisition of a nuclear force will not automatically deter the great nuclear powers and may increase the danger by encouraging the spread of nuclear weapons to neighbours.

---

[44] See, for example, Gallois, Pierre, *The Balance of Terror: Strategy for the Nuclear Age* (Boston: Houghton Mifflin, 1961).

[45] Speech to the American Press Club, Paris, 5 March, 1963, p. 2 (typescript).

The main reward for promising not to manufacture or receive nuclear weapons is a corresponding promise by some other countries, backed by a system which can provide effective early warning if promises should be broken, and by the prospect of avoiding or reducing an increasingly menacing disorder. Such compensatory promises do not take care of threats from those states which have already acquired nuclear weapons. Alliances with firm guarantees will be essential, since in any case a few weapons may by no means suffice for deterrence. This is true in particular of the problem of deterring attack by super-powers on small powers or the coercion of such powers by super-powers.

Strategic and Tactical Warning Time

A system of safeguards—whether bilateral or multilateral—can at most sound an alarm. But what events do we want to be warned about? There are two possibilities. The first is that a government which has been a long way from being able to manufacture nuclear weapons is now within a short distance of that capacity; this is a kind of strategic warning. The second is that a government has " diverted " special nuclear material in violation of an international agreement; this would be a kind of tactical warning.

These two events differ. The first is more fundamental; it can occur even if the second does not—if agreements do not preclude drastically shortening the time to make a nuclear explosive. Strategic indications now abound and their implication is that tactical warning is in some cases dwindling to days or hours.

MUF, " MAF ", and IAEA: The proximate goal of safeguards has been to prevent " diversion ". For this reason, the inspectorate of IAEA, and possibly the inspectors of an exporting country in a purely bilateral arrangement, fix their gaze on the " material unaccounted for " (MUF) and the " limits of error in the material unaccounted for " (LEMUF). The safeguards of the Non-Proliferation Treaty, for example, consist essentially in verification of accounting records to see whether the quantity of the material unaccounted for falls within reasonable limits. American officials contend that the limits of error can be reduced to less than half of 1 per cent.; officials of other governments think they cannot be reduced much below 1 per cent. From the standpoint of diversion, it is this very small percentage which is problematic.

However, from the standpoint of preventing the future spread of capacities to make nuclear explosives with a short delay, it is trends in the larger fraction of materials accounted for (" MAF ") which are likely to be decisive. For various reasons, the basic data on the special nuclear materials which are adequately accounted for, owned and controlled by national governments, possibly stocked within their territory and tidily reported at regular intervals in accordance with bilateral or trilateral agreements, are not assembled and reported. They are not what " safeguards " are about, if " safeguards " are taken to mean merely inspection to detect violations. Safeguards have to do with material which conceivably may have been diverted.

Moreover, the IAEA system of safeguards [46] specifically prohibits disclosure by any member of the IAEA staff—except to the director-general and other members of the IAEA designated by the director-general—of any "confidential information" acquired while carrying out agency safeguards procedures.

Paragraph 14 of the basic document of the "safeguards system" of the IAEA states:

The Agency shall not publish or communicate to any State, organization or person any information obtained by it in connection with the implementation of safeguards, except that:

(a) Specific information relating to such implementation in a State may be given to the Board and to such Agency staff members as require such knowledge by reason of their official duties in connection with safeguards, but only to the extent necessary for the Agency to fulfill its safeguards responsibilities;

(b) Summarized lists of items being safeguarded by the Agency may be published upon decision of the Board; and

(c) Additional information may be published upon decision of the Board and if all States directly concerned agree.

Even exception " (b) " which the IAEA Board might make is limited to lists of items being safeguarded. There is no exception which permits the reporting of the quantities of such items.

The IAEA interprets its mandate as forbidding any report on the size, physical state, and isotopic and chemical composition of any stocks of special nuclear material present and accounted for, not to say those missing and unaccounted for at specific facilities, except to a very narrowly circumscribed set of staff-members of IAEA. In practice, this has meant that up to now [47] the data on such material accounted for have not been assembled and published, even in the form of frequency distributions of the number of countries with various amounts of material of varied type. Still less has there been any analysis of trends in such stocks accounted for. Attention, in short, has been on the half of 1 per cent. which might be unaccounted for and not on the 99·5 per cent. or so which might present the most important problem.

Indeed, the state of public information on these civilian stocks appears to be much worse than that on the supposedly secret and highly guarded stocks of missiles and military aircraft, etc., which form the regular subject of annual military balances by the International Institute of Strategic Studies and similar organisations. This seems strange. It has its origins in the concern of various competing nuclear exporters for the protection

---

[46] International Atomic Energy Agency, " The Agency's Safeguards System, 1965, as Provisionally Extended in 1966 ", INFCIRC/66/rev. 1, 1967, IB, para. 13.

[47] That is, up to the date of writing. Legislation passed by the United States House of Representatives and under consideration by the Senate in October 1977 (The Nuclear Antiproliferation Act of 1977, HR8638) would address this matter by requiring such countries engaged in nuclear commerce with the United States to consent to the release by the IAEA of such data, by stating that the peaceful nuclear activities to which the United States subscribes ought to be such as " to permit warning time of diversion comparable to that which has traditionally obtained in the low enriched fuel cycle without reprocessing ", and by making " timely warning " the " foremost " factor in determining whether or not to permit the reprocessing of nuclear material suplied by the United States.

of commercial and industrial secrets. Paragraph 13 of the basic document of the " safeguards system " states:

In implementing safeguards, the Agency shall take every precaution to protect commercial and industrial secrets. No member of the Agency's staff shall disclose, except to the Director General and to such other members of the staff as the Director General may authorize to have such information by reason of their official duties in connection with safeguards, any commercial or industrial secret or any other confidential information coming to his knowledge by reason of the implementation of safeguards by the Agency.

In the course of checking material unaccounted for an inspector must necessarily come by and use information regarding the material which is accounted for. The rules suggest a presumption that any such information is confidential and not to be disclosed. It seems quite far-fetched, however, to suppose that information about inventories of separated plutonium and highly enriched uranium would reveal any commercial or industrial secrets of significance to competitors. In fact, scraps of such information can be obtained from publicly accessible sources—where, for example, the source materials have been obtained under licence from the United States Government. It is possible to determine that two years ago Spain had an inventory of between 60 and 70 kilograms of separated plutonium at Windscale, England. This fact is indeed of interest to anyone concerned with the possibility of the future uses of plutonium in nuclear weapons, but it reveals no secret of commercial importance about the design of reactors or similar matters.

Apparently a sweeping restriction, born of fond hopes for preserving competitive advantage in some ingenious trick of reactor design, has resulted in more confidentiality than is needed and a failure to disclose information of great public interest and little industrial importance. Whatever the genesis of these constraints on disclosure of information about stocks, the effects are quite perverse. It means that where an exporting country has substituted for a bilateral agreement a trilateral one between itself, an importing country and the IAEA, it is prevented from receiving information on a possible violation. It can receive only information of confirmed violations, after a substantial delay, and then only indirectly. This reduces the warning time about possible violations and restricts the possibilities that the exporting country might, by diplomatic means, influence the potential violator to reverse its course. The most important effects, however, have to do not with warning of violation, but with warning of the approach towards a capacity to produce nuclear weapons—without violation. For the formation of timely and precisely adapted actions to inhibit the spread of nuclear weapons, the regular publication and analysis of trends in the stocks of fissionable material accounted for would be extremely important.

The Effective Application of Safeguards to Assure Timely Warning

The proposals made in 1945 and 1946 by the pioneers of the discussion of international control of nuclear energy may have been seriously at fault so far as the technical means for assuring control were concerned. They were a good deal clearer and more explicit about the purpose of such controls, however, than the vast bulk of the technical discussions which ensued in the 1950s and the 1960s. Only in recent years has something like the original clarity about the purpose of safeguards been restored. Much of the discussion of techniques of inspection in the 1960s continued to be influenced by the initial technical flaws, particularly the hope—still persisting in implicit form—of " denaturing ".

It is hard to understand how so much effort could have been spent in the 1960s on elaborate schemes for " safeguarding " reprocessing plants and plants for manufacturing mixed plutonium and uranium dioxide fuel, both of which would result in stocks of separated plutonium quickly use-able in bombs. It may be explained by the persistence of the implicit belief that plutonium separated chemically from fuel used in power reactors would be safe because it was contaminated by plutonium-240 and -242; the chemical separation barrier could be safely removed, so to speak, because the even higher isotopic separation barrier to an explosive would remain. Or it may be the result of attention so concentrated on the technical means of inspection to detect violation as to lose sight of the purpose of the agreement on safeguards and inspection. The purpose of safeguards is to assure timely warning which would deter the manufacture of nuclear explosives or permit appropriate national or international responses to frustrate it.

A concentration on the means to some proximate end, such as " detection ", is characteristic of the behaviour of large organisations. It is less characteristic for them to be concerned with whether the means to that proximate end are compatible with more remote objectives, such as the assurance that no activity will be undertaken which would prevent timely response. It is still rarer to find a searching inquiry as to whether the remote objectives are compatible with each other: whether, for example, the assurance of early warning is compatible with the alleged necessity of " closing " the fuel cycle by recycling plutonium in order to accelerate and enlarge commerce in nuclear energy throughout the world.

The belief that it is essential to close the fuel cycle provides a powerful impulse to hold fast to the myth of denaturing plutonium. That so many high officials have continued to cling to this conviction or at least to the hope that diluting plutonium-239 with plutonium-240 or -242 would make plutonium-239 safe is remarkable, given the doubts which were expressed about such a proposal even before the Acheson-Lilienthal report. It testifies perhaps to our wish that plutonium used in a power reactor should be safe rather than our firm conviction that it is. Interposing the high barrier of isotopic separation between the use of plutonium for civil and military ends, if it could be arranged, would provide early warning even with the commercial methods of isotopic separation prevailing today. Early warning, it was understood at the outset, was part of the essence of safeguards. The feasibility of such dilution of plutonium was called into question a year before the Acheson-Lilienthal report by Dr. Glenn Seaborg, during the writing of the Franck report of which he was co-author. The Franck report, which was dated 11 June, 1945, was concerned most immediately with the decision to drop the bomb on Japan, but it also suggested that a " compulsory denaturation of pure fissionable isotopes may be agreed upon by diluting them, after production, with suitable isotopes to make them useless for military purposes, while retaining their usefulness for power engines ". Dr. Seaborg, commenting on early drafts of the report,[48] wrote " can't denature 49 by dilution with stable isotopes " [49] (" 49 " was the wartime code for plutonium, the atomic number of which is 94.) Yet the Franck report was issued without change, presaging our persistent ambivalence.

The early proponents of the idea of making plutonium safe by contaminating it with stable higher isotopes were ambivalent and troubled by its substantial technical shortcomings. Nonetheless, they clearly and unequivocally grasped an essential fact of international safeguards which is that a safeguard involves more than simply detecting a violation of an agreement. Though reasserted from time to time, this has a way of getting lost in the middle reaches of international and national bureaucracies. A system of inspection, to deserve the reassuring name " safeguard ", requires that the approach by one government towards the production of a nuclear weapon be detected in time for other governments to do something about it.

Leo Szilard, who discussed the possibility of denaturing even earlier than the Franck report and rather more cautiously—he said, " the possibility would have to be scrutinised carefully "—was quite clear on the critical need for timely detection :

A system of controls could be considered successful only if we could count on a period of grace in case the controls were denounced or obstructed by one of the major powers. This means that the system would have to be of such a nature that at least one or two years would elapse between the time the nations begin to convert their installations toward the purpose of manufacturing atomic bombs, and the time such bombs become available in quantity.[50]

The hopes for a denaturing of plutonium which would compel isotopic separation were disappointed. However, chemical separation remains as a barrier which takes a substantial, if smaller, amount of time to surmount, provided spent uranium fuel is controlled. Any interpretation of safeguards which removes this last barrier of chemical separation for plutonium and leaves practically no warning time should be recognised as abandoning the essential purpose of safeguards.

If the critical time to make an explosive is allowed to shrink to a few

[48] Now available in the James Franck papers at the Regenstein Library of the University of Chicago.

[49] Dr. Seaborg, in a letter of 10 March, 1976, stated: " My second written comment that ' can't denature 49 by dilution with stable isotopes ' certainly refers to my view that isotope Pu-240 was being greatly over-rated as a safeguard against nuclear proliferation. I remember feeling that way the next year when the Lilienthal Board suggested the possibility of denaturing fissionable Pu-239. (Denaturing fissionable U-235 and U-233 is, on the other hand, feasible due to the availability of more stable non-fissionable U-238.) I don't recall why my written comment didn't lead to an alteration in the final version."

[50] " Atomic Bombs and the Post-War Position of the United States in the World ", March 1945. Typescript available in the office of the ERDA historian.

weeks, days, or hours, there will not be enough time for political or military action. For one thing, deciding on appropriate political or military responses would take time, and political responses in themselves might be very time-consuming. For another, the warning would not begin with the first moment of diversion. The first indication of violation an inspector might have, even if he were on the scene, might be some denial of access, ostensibly perhaps because of a dangerous radioactive spill. This might be only the beginning of a long sequence of events, restricted at first to the inspected government and the inspector. The IAEA safeguards might assure quick detection for a violation shortly after it takes place, but it cannot assure timely detection well before the assembly of bombs unless agreed arrangements between importers and exporters keep a sufficiently long critical time between violation and assembly.

Both the IAEA and the United States government have reiterated the essential point, that the purpose of safeguards—and therefore the objective which would have to be achieved for the application of safeguards to be " effective "—is to assure timely detection and the deterrence of diversion through the risk of early detection. That point was made in a report transmitted by President Ford to Congress. He said: " The international safeguard system deters diversion by the threat of early detection of diversions should they occur at the national level and by the political consequences resulting from reporting of diversion to the international community." [51] General Starbird repeated the point a short time afterwards: " IAEA safeguards provide for the timely detection of diversion . . . and the deterrence . . . of such diversion by threat of early detection." [52] His assertion faithfully reproduces the objective of safeguards as clearly defined by the safeguards committee of the IAEA: " The timely detection of diversion of significant quantities of nuclear material from peaceful nuclear activities to the manufacture of nuclear explosive devices or for purposes unknown, and deterrence of such diversion by risk of early detection." [53] The essentiality of timely warning has been reaffirmed by President Carter and is embodied in the bill drafted by the House of Representatives.[54]

The point of it all is sometimes lost. But as the inspector general of IAEA suggests, " The thirty years history of safeguards is one of continuous development and there is no reason why its techniques and concepts should not undergo many further changes." [55] There is every reason to believe that further evolution should clarify and make more precise the design of safeguards to enable them to serve their essential purpose.

[51] *Laws and Regulations Governing Nuclear Exports and Domestic and International Nuclear Safeguards,* Message from the President of the United States, 6 May, 1975 (Washington, D.C.: U.S. Government Printing Office, 1975), p. 35.

[52] Statement by the Assistant Administrator for National Security of the Energy Research and Development Agency, 29 January, 1976. U.S. Senate, Committee on Government Operation, *Hearings on the Export Reorganization Act of 1976* (Washington, D.C.: U.S. Government Printing Office, 1976), p. 408.

[53] See IAEA INFCIRC/153 (1971); and Sanders, B. and Rometsch, R., " Safeguards Against Use of Nuclear Materials for Weapons ", *Nuclear Engineering International*, XX, 234 (September 1975), p. 683.

[54] See footnote 47 above.

[55] Sanders, B. and Rometsch, R., *loc. cit.*

ECONOMIC ASPECTS OF NUCLEAR PROLIFERATION

KNOWLEDGE of the economics of nuclear power can assist in improving policy on proliferation in at least three ways. First, economic analysis can improve projections of the growth in capacity for development of nuclear weapons: the more that economics favour nuclear power relative to fossil fuel power, and in particular the more they favour those forms of nuclear power which make concentrated fissile material accessible, the more rapid would be the expected growth in weapons capability. In spite of suppliers' promotional efforts, few nuclear plants were sold to less developed countries during the 1960s. The high costs of nuclear power then were obviously very influential, since prestige, independence, and national security all seemed to favour nuclear power.

Second, economics are essential in determining whether restraints on civilian activities to avoid the spread of weapons will create serious scarcities of nuclear fuel. A shortage of uranium is not simply a matter of geology. Sea water and the top layer of the earth's crust contain enough uranium to fuel existing and planned nuclear reactors for many thousands of years. What fraction of this uranium is worth the cost of extraction is an economic question.

Third, economics can aid in formulating nuclear export policies. There are not two separate types of atom—one " peaceful " and the other " military "—but some civilian nuclear activities are more dangerous than others. These can be discouraged or banned rather than promoted. In fact, the United States has not transferred certain technologies with civilian applications, especially enrichment technology. Assessing the potential for proliferation of a specific technology is one aspect of a prudent export policy. The dangers from nuclear exports need to be compared with their potential economic benefits.

The supposed benefits of alternative nuclear programmes to importing as well as exporting countries need to be examined critically. It is the small or developing countries which now appear most likely to move towards a decision to make nuclear bombs in the next decade. We should not assume uncritically that any restraint on nuclear activity would involve an immoral denial of vital economic benefits to such countries. Not all civilian nuclear activities have equal economic benefits. Some may even be economically inferior to conventional alternatives. The essence of policy choice is striking the proper balance between economic desirability and military danger. To ignore economic benefits implies that all civilian nuclear activities should be opposed. To ignore military danger implies support for all activities, no matter how dangerous.

Our analysis shows that the disorder attending the rapid spread of nuclear weapons would endanger importing as well as exporting countries. Even regional antagonisms would become much more precarious and lethal. But are the promised economic advantages large enough to offset such dangers? Evidence presented in the first part of this chapter suggests that they are not.

The latter part of this chapter treats in depth a central issue in the current nuclear debate: whether or not spent nuclear fuel should be reprocessed to recover the contained plutonium and uranium for further use. This issue represents a clear example of the importance of economics in formulating nuclear policy. The military dangers of plutonium fuels are clear and substantial: within weeks or even days, these supposedly civilian materials could be converted by a country into material for nuclear bombs. In contrast, the analyses presented here show that near-term introduction of these fuels would produce, at best, insignificant economic benefits. This marked imbalance of risks and benefits provides a strong argument for deferring decision on the introduction of plutonium into the nuclear economy.

Exports and the Economics of Nuclear Power in Less Industrialised Countries [1]   .

Inflated projections of growth in nuclear capacity: Rapid expansion of nuclear power in less developed countries has often been forecast, beginning with the "atoms for peace" predictions in 1954. These predictions have been unfulfilled, but the large rise in fossil fuel prices dating from the Middle East October War of 1973 has created a widespread belief that the " take-off " of nuclear power lies just ahead. The Organisation for Economic Cooperation and Development (OECD) and the International Atomic Energy Agency (IAEA), for example, have recently predicted rapid

TABLE I

Nuclear Power Plants in Less Industrialised Countries
(Megawatts of Electric Capacity, 30 June, 1976)

Country	Commercial Operation	Under Construction	On Order	Total Initiated Capacity
Argentina	319	600		919
Brazil		626	2 × 1245	3,116
India	3 × 200	200 + 2 × 220	2 × 220	1,680
Iran			2 × 1200 + 2 × 900	4,200
South Korea		564	605 + 629	1,798
Mexico			2 × 660	1,320
Pakistan	125			125
Philippines			2 × 626	1,252
Taiwan		2 × 604 + 951	951 + 2 × 907	4,924
Totals	1,044	4,589	13,701	19,960

SOURCE: *Nuclear News*, XIX, 10 (August 1976), pp. 66–72.

[1] All countries outside Europe and North America except Japan, China, the Soviet Union, Israel and South Africa.

expansion of nuclear power capacities in less developed countries in the next decade. These predictions have apparently been accepted without significant question.

Our study of the projections by the IAEA and OECD suggests that the optimists have greatly overestimated the future growth of nuclear power. A very small amount of nuclear capacity, 1,044 MW (megawatts), was actually operating in 1976 (Table I). This is equivalent to a single plant of the typical size now being built in the United States and Europe. The "total initiated capacity"—plants operating, under construction, and on order—provides a useful yardstick for estimating installed nuclear capacity in 1985. This is so because it typically takes about 8 to 10 years from the date of order to the date of initial operation. Although this period may sometimes be shorter, experience strongly indicates that most of the plants operating in 1985 will have been initiated by now. But, for the projections of the OECD and IAEA to be fulfilled, installed capacity in 1985 would have to exceed considerably the amount now initiated.

The OECD has projected a nuclear capacity of 25,000 MW in 1985 for the less developed countries in their organisation.[2] The amount of capacity now initiated by these countries is 15,760 MW,[3] or 60 per cent. of the amount projected.[4]

The IAEA has provided estimates for 14 individual countries. The total projected amount in these countries studied by the IAEA is 5·5 times the amount now initiated (Table II). The possibility of having this amount in operation seems extremely remote.[5]

A look at the methods used to arrive at these projections reveals a number of questionable assumptions. First, high growth rates: for the period 1981 to 1985, the average growth rate in the 14 countries was assumed to be 9·3 per cent. per year.[6] Although not specified in the report, the method used implies that a very similar growth was assumed for 1973–80. This growth rate, compounded for 12 years, implies a total growth of 200 per cent. for the period. This rate seems unrealistically high. In fact, since the oil embargo and price jump of 1973, growth of electricity consumption in Europe and the United States has been quite low. Although data were not available to use for less developed countries, the same factors which slowed down growth of electricity consumption in the industrialised countries were at work in the less developed countries,

[2] *Energy Prospects to 1985*, Vols. I and II (Paris: OECD, 1975).

[3] " Total initiated capacity " from Table I for all countries except Iran, which is not a member of OECD.

[4] The OECD does not provide a country-by-country listing nor any details on the method used to arrive at the projections.

[5] The IAEA in its annual report of 1975 recognised that their projection in 1974 for 1980 was much too high. The annual report of 1975 estimates worldwide installed nuclear capacity at 220,000 to 250,000 MW, about 30 per cent. less than their estimate of 1974. See " IAEA: Move to Nuclear Power Slower Than Expected ", *Nuclear News*, XVIII, 11 (September 1975), p. 66. Similar reductions in estimates for 1985 can be expected as actual construction plans become known to the IAEA.

[6] See " Total High Forecast ", Table V–1, of the *Market Survey General Report* (Vienna: IAEA, September 1973), p. 21. The " High Forecast " was used in the *Market Survey* (1974 edition) which brings the original study up to date in light of changes in the price of oil and nuclear power costs which had occurred by early 1974.

TABLE II

Initiated versus Projected Nuclear Capacity

Country	Initiated [a] as of 30 June, 1975	IAEA Projections for 1985
	Megawatts	
Argentina	919	3,500
Bangladesh	0	1,300
Chile	0	600
Egypt	0	1,200
Greece	0	2,000
Jamaica	0	600
South Korea	1,798	4,200
Mexico	1,320	7,800
Pakistan	125	1,325
Philippines	1,252	1,200
Singapore	0	1,650
Thailand	0	1,400
Turkey	0	1,200
Yugoslavia	0	2,800
Totals	5,414	30,775

" Initiated " is defined as capacity currently operating, under construction, or on order.
SOURCE: International Atomic Energy Agency, *Market Survey for Nuclear Power in Developing Countries* (Vienna: IAEA, 1974).

perhaps intensified. Second, the relation of GNP to electricity: in the IAEA study, the growth in demand for electricity was derived from an analysis implying a fixed relation between growth in *per capita* GNP and growth in *per capita* demand for electricity. There are many theoretical reasons for questioning this assumption, but even its statistical basis appears defective. The relation was derived from cross-country plots of electricity use *per capita* versus GNP *per capita*—whereas our interest is in how changes in the first quantity are related to changes in the second quantity. The correlation between these changes is not at all consistent, especially at the low levels of *per capita* GNP and electricity generation appropriate to less developed countries.[7] Third, relative cost of fossil and nuclear power: the IAEA study omits coal as an alternative to nuclear power, but coal is much cheaper than oil as a source of fossil energy. Omitting coal as an alternative to oil creates a serious bias in the IAEA study.

Even more fundamental is the inadequate treatment of the uncertainties of the economics of nuclear power. The analysis in the IAEA report of 1975 shows such a marked cost advantage for nuclear power that all future installed capacity—except one 100-MW plant in Jamaica—is nuclear. That no such overwhelming cost advantage for nuclear power exists is

[7] This is apparent in *Market Survey General Report* . . . Figure F-3, p. F-5.

suggested by recent developments in Korea. Four, out of nine, nuclear plants have been dropped from the Korean electricity-generating plan, and at the same time, the Korean Electric Company is negotiating orders on two 300-MW combined-cycle units and 600 MW of oil-fired capacity.[8] The reasons cited are that the fossil fuel plants are cheaper to begin with, easier to finance, can be built faster, and are less controversial than nuclear units. The possibility that nuclear power may often be more expensive than the alternatives is explored in greater detail below.

Important uncertainties in the cost of nuclear and fossil power in less developed countries: Official estimates of the costs of nuclear versus fossil power in the less developed countries have a number of deficiencies. First, capital costs are based primarily on experience of construction in the United States and Europe, adjusted for differences in wage rates and equipment costs. This method leads to estimates of costs which are lower for less developed countries than for corresponding plants built in the United States. Actual experience, however, shows the opposite to be true. Because less developed countries are generally far from the suppliers of nuclear components and lack the highly-skilled workers required for nuclear construction, the cost of building nuclear plants in less developed countries are turning out to be higher than in industrialised countries.[9] Second, in calculating electricity costs, official estimates assume operating rates of reliability based on engineering goals rather than actual experience. These goals have not been met in the United States and the increased difficulty of repairs suggests that they are likely to be even lower in the less developed countries. Third, they ignore the historical experience of capital cost increases, which have been significantly greater for nuclear than for fossil plants. Fourth, they use interest rates which are appropriate to industrialised countries but which underestimate the rates in less developed countries. Fifth they use high prices for coal and oil but ignore the higher costs of the nuclear fuel cycle, which are already evident. All of these deficiencies favour nuclear power.

Estimates which show rapid nuclear growth generally have been prepared by agencies and experts with vested interests in the success of nuclear power. " Reasonable guesses " are often required in cost estimation; when the estimator has a marked bias, the cumulative effect of the bias can be very large. In contrast to the confident, single-valued estimates presented in the OECD and IAEA studies, experts speaking privately often emphasise the massive uncertainties which characterise future estimates. A recent review of nuclear and fossil power costs in the United States found many uncertainties.[10] Figure 1, which shows the possible effects of these uncertainties, reveals that there is a much wider range of uncertainty for nuclear power costs than for fossil power costs. The authors note that: " As a consequence, we find that nuclear power may turn out to produce

---

[8] " South Korea Drastically Cuts Back Nuclear Program; Finance the Problem ", *Nucleonics Week*, XVII, 27 (1 July, 1976), p. 2.

[9] " Nuclear Export Market Now Largely Restricted to Third-World Nations ", *Nucleonics Week*, XVII, 36 (2 September, 1976), p. 5.

[10] Holiday, D. and Taylor, V., *The Uncertain Future of Nuclear Power* (Santa Monica : California Seminar on Arms Control, August 1975).

electricity that costs several times more than coal power (if the uncertainties are resolved unfavourably for nuclear power) or slightly less (if the uncertainties are resolved favourably)." [11]

FIGURE 1

The Range of Uncertainty in Future Costs of Generating Electricity by Alternative Methods (1974 Dollars)

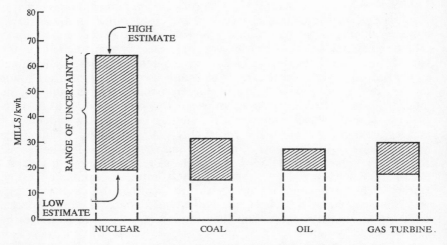

SOURCE: Holliday, D. and Taylor, V., " The Uncertain Future of Nuclear Power " (Santa Monica: California Seminar on Arms Control, August 1975), Table I, pp. 10–11.

The deficiencies in methodology used by the IAEA to compare the economics of nuclear and fossil power in the less developed countries are sufficiently serious to make the comparison of little value. To improve policy, better economic analyses will be required.

The limited potential benefits of nuclear power: The promised benefits of nuclear power for the less developed countries have proved to be elusive. Somehow they are always just ahead, but never quite in hand. If oil prices remain at current levels, nuclear power may well prove to be the cheapest source of power for some countries, but it is not going to usher in an economic millenium. At best, it may bring slightly cheaper electricity to some countries. For the most part, the annual savings *per capita* in third world countries will be measured in pennies. Even in cases where nuclear electricity is cheaper than fossil power, the cost advantage seems unlikely to be more than a few mills per kwh. At typical operating rates, a cost advantage of 5 mills per kwh becomes an annual saving of $25 million for each 1,000 MW of installed nuclear capacity. To gain this annual saving, smaller countries would have to invest about $500 million more for 1,000 MW of nuclear power than for equal fossil capacity.

[11] *Ibid.*, pp. 1–2.

The potential savings from nuclear power can be translated into *per capita* savings. In most countries, these benefits, if they materialise, would amount to less than one dollar per person per year. For Argentina, Brazil, Taiwan, India, Israel, Korea, Mexico, Pakistan, Philippines, South Africa, and Thailand—assuming that the optimistic, perhaps excessive, nuclear growth projections of the OECD were met—the average *per capita* savings from nuclear power—assuming a 5 mills kwh advantage—would be 64 cents in 1985 and $1.28 in 1990.

Subsidies to encourage nuclear exports: The United States Export-Import Bank in 1974 alone authorised direct loans of $439 million and guaranteed another $162 million for nuclear power plants and training centres abroad which purchased equipment made in the United States. From the inception of its programme of subsidies of nuclear exports, the Bank has directly loaned $2.47 billion and guaranteed an additional $931 million. In addition, the Agency for International Development has provided a highly-subsidised loan of $72 million for a reactor at Tarapur, India, and transferred at least an additional $30 million to underdeveloped countries for research reactors, support of laboratories, and nuclear technical assistance.

In December 1963, the United States authorised a loan of $71·8 million to India for the Tarapur nuclear power station. Interest on the loan was three fourths of 1 per cent. during both a 10-year period of grace and a 30-year repayment period. This was a very low interest rate, even for 1963, and the maturity was exceptionally long. In effect, a large part of this loan amounted to a gift. The extent of the gift depends upon the terms which India would have had to pay in the full market. For example, if an Indian utility going into the market for a comparable loan would have had to pay 10 per cent., the discounted cost to India of the stream of loan payments to the United States would have been only $13 million. The remaining $58·7 million or 82 per cent. of the amount of the loan would represent a gift from the United States.

The Export-Import Bank reports that it has financed about 60 per cent. of the cost to the United States, and about 25 per cent. of total costs for the nuclear power projects in which it has participated. The Export-Import Bank has also given financial guarantees to an additional 10 per cent. of total project costs. Its interest rates are well below market rates, and its repayment terms are far more generous than those for normal commercial loans. In addition, in order to reduce the risk to private lenders, the Export-Import Bank, at least sometimes, gets repaid from the final set of payments. This, of course, increases the effective risk to the Bank. Subsidies for nuclear exports make up a large part of the United States Export-Import Bank's business. The president of the Bank has estimated future demand for nuclear export loans of $750 million a year out of an authorised annual total of $3·4 billion. He has stated that " nuclear power rates a very high priority ".[12]

[12] Crittenden, Ann, " Surge in Nuclear Exports Spurs Drives for Controls ", *The New York Times*, 17 August, 1975.

The United States is not alone in subsidising nuclear exports. The governments of Germany, France, and Canada also provide favourable financing to facilitate sales of reactors. The German government has been instrumental in arranging loans of $1.7 billion to Brazil to finance the sale of two reactors by Kraftwerk Union (KWU), the large German nuclear manufacturer.[13] A government bank is supplying over 40 per cent. of the loan, some (unspecified) part of which will be at 7·25 per cent. Other German private banks and their European branches are supplying the remainder at rates " in line with market conditions ". Most of these loans are for 20 years, a term which no commercial bank would provide without governmental involvement. The extent of the loan subsidies provided by nuclear-supplier countries is indicated by an analysis in 1969 of J. H. Cha of the Korean Atomic Energy Research Institute. He estimated an interest rate of 6·5 per cent. per year on a foreign, subsidised loan on a Korean nuclear plant and a rate of 16 per cent. per year on domestically raised funds for the same plant.[14]

The argument used in all countries to justify subsidies is, " If we don't provide them, other governments will, and we will lose the business ". Adam Smith pointed out that if other countries want to tax the general population for the benefit of a few suppliers, that is their problem, but it is not a persuasive argument for one's own country committing the same folly. This seems especially true for nuclear exports, given their military potential. Further, with a floating exchange rate, specific exports, such as nuclear power plants, are not required to bring about a satisfactory balance of payments.

The United States and other supplier nations are not merely providing the technical means for creating the option for nuclear military forces in other countries; they are helping to pay for this option with their subsidies.

The Economics of Plutonium and Uranium

On 28 October, 1976, the President of the United States announced an historic change of policy when he stated: " I have decided that the United States should no longer regard reprocessing of used nuclear fuel to produce plutonium as a necessary and inevitable step in the nuclear fuel cycle, and that we should pursue reprocessing and recycling in the future only if they are found to be consistent with our international objectives." [15] He stated the central reason for his decision: " Unfortunately—and this is the root of the problem—the same plutonium produced in nuclear plants can, when chemically separated, also be used to make nuclear explosives." President Carter has moved a step further by declaring that the United States " will defer indefinitely commercial reprocessing and recycling of plutonium produced in the U.S. nuclear power program ".[16] For the first

[13] " Brazil Is Getting Large and Long-Term Financing ", *Nucleonics Week*, XVII, 30 (22 July, 1976), p. 13.

[14] Cha, J. H., " Estimate of Nuclear Power Costs in Korea ", Proceedings of an IAEA Symposium in Istanbul, 20–24 October, 1969, IAEA-SM-126/40, 1970.

[15] " Statement of the President on Nuclear Policy ", Office of the White House Press Secretary, 28 October, 1976.

[16] " Statement of the President on Nuclear Power Policy ", Office of the White House Press Secretary, 7 April, 1977.

time, keeping plutonium out of general circulation in the nuclear economy appears to be a realistic possibility. But, at this stage, it is only a possibility. Translating it into a reality will be no easy task.

For the past 30 years, the programmes and actions of the United States Atomic Energy Commission, its successor, and its counterparts abroad have supported development of technology to utilise plutonium as fuel in light water and breeder reactors. The use of plutonium has been widely promoted as a potential solution to the exhaustion of fossil energy resources. All these activities have created interests and opinions which will work against the goal of keeping plutonium out of circulation.

Deferring use of plutonium fuels: The President's decision was strongly influenced by an appreciation of the great changes which have taken place in the expected economics of reprocessing and use of plutonium. Only a few years ago, reprocessing was generally assumed to promise significant economic gains, but as the full complexity of the task has become appreciated, the previously expected gains have become increasingly doubtful.

Careful analysis of the latest available information on fuel cycle costs leads to the following conclusions: First, it is uncertain at this time whether introduction in the near future of reprocessing and a plutonium fuel cycle—commonly and misleadingly termed fuel " recycle "—would lead to net economic gains or losses, although losses seem more likely, especially in less developed countries where the small scale of operation would heavily penalise the economics. Second, even if uncertainties of the fuel cycle were resolved favourably, its impact on the cost of nuclear power would be insignificant—a reduction of less than 1 per cent. in the average delivered costs of nuclear-generated electricity.

The " costs " of proliferation, environmental damage and terrorism which might result from widespread use of plutonium fuels, although difficult to quantify, appear large enough to justify foregoing a large benefit in order to avoid them. But, the potential benefits of reprocessing are uncertain at best, and, at the upper limit, not significant in comparison with the total costs of nuclear power.

These considerations argue for a strong effort to gain worldwide agreement to defer general reprocessing of spent fuel to recover fissile products until such time as the use of plutonium fuels in light water reactors can be confidently shown to yield economic benefits sufficiently great to compensate for the large non-monetary costs, or until the viability and safety of the breeder reactor as an important commercial source of power have been demonstrated.

Given the low upper-limit of the benefits from use of plutonium fuels in light water reactors (LWRs), the second possibility seems far more likely to provide a justification for reprocessing. But when, and even if, the plutonium breeder becomes competitive with thermal reactors in terms of economics and safety is very uncertain. The United States has recently reduced drastically funds for development of the breeder reactor, making it unlikely that any significant number of breeders will be installed in the United States in this century. Although European development programmes have progressed further than that of the United States, important technical problems remain. Nor is it certain that a breeder will emerge which is competitive with the generation of LWRs which is then current. When this might occur is extremely uncertain, but it will clearly not be before the 1990s. Thus the breeder programme provides no short-term justification for reprocessing. If commercial production of breeders stretches far enough into the future, developments could take place which would obviate the need for it. If so, a decision to defer reprocessing could result in entirely avoiding the risks inherent in the widespread use of plutonium fuels.

If use of plutonium fuel is deferred until such time as substantial bene-fits can be confidently predicted, the most which could be lost would be the small benefits which might have accrued during the period of deferral. The major, long-term benefits will still be obtainable. By contrast, if countries decide now to go ahead with plutonium use and it turns out to have been a mistake, the costs of proliferation will already have been borne, with little chance of reversal. Moreover, the nuclear energy sector will have burdened itself, and everyone else, with political, environmental, and economic problems worse than any it would otherwise face.

The decision on use of plutonium fuel in LWRs is viewed by many nuclear proponents as a crucial battle in the war against the anti-nuclear faction, but analysis shows that the economics of nuclear power will be little affected by the outcome. Engaging in this battle will require the proponents to use inordinate amounts of political capital and technical talent. Yet, the major result of a victory seems likely to be a further substantial increase in opposition to nuclear power. Given that public acceptability appears to be a primary obstacle to growth of nuclear energy, friends of nuclear power would do well to consider thoughtfully whether continued advocacy of immediate reprocessing serves well their own objectives.

Although the United States provides the central focus for analysis here, the conclusions which emerge apply with at least equal force to other coun-tries. All countries share a common concern in seeing that civilian nuclear power does not bring additional countries dangerously close to nuclear bombs, as surely would happen if plutonium were to become widespread. The threat to national security would be far more direct for some other countries than for the United States, and the economic messages are as applicable to other countries as to the United States. For smaller countries limited to domestic markets—i.e. no exports or consortiums—the economics would be less favourable than in the United States because of higher costs per unit associated with the necessarily smaller scale of oper-ations. Germany and the United Kingdom are considering investments in reprocessing more substantial than those now under way in the United States.[17] Our analysis shows that reprocessing seems more likely to com-

[17] Britain is contemplating a billion dollar expansion of its existing Windscale reprocess-ing plant, and there are German plans for construction of a 1,400-ton per year reprocessing plant for operation in 1988–89, at an estimated cost of $1.6 billion, at present prices: see *Nucleonics Week*, XVIII, 10 (March, 1977), p. 7. The German plant would be over 50 per cent. more expensive than assumed in the analysis by Pan Heuristics of reprocessing economics, an analysis which shows that reprocessing would be very unprofitable at anticipated uranium prices.

plicate than to simplify the task of managing radioactive wastes. Thus, those countries considering large investments in reprocessing may wish to consider especially carefully the large risks and limited potential gains associated with such investments.

The likelihood of sufficient low-cost uranium: The economic desirability of plutonium fuels depends to an important extent on the availability and cost of uranium fuels. The history of the uranium industry and present evidence strongly suggest that large amounts of uranium will be forthcoming at relatively low prices. Based on present technology for producing uranium, the uranium requirements of the United States up to the year 2000 appear capable of being met from its own resources which can be discovered, mined, and milled for less than $20 per pound of uranium oxide (U_3O_8) in 1975 dollars. With uranium at such prices, the use of plutonium fuels makes no economic sense. For the entire non-communist world, the supply of uranium appears likely to be more favourable. Further, continuation of past trends towards lower production costs—caused by sustained technological improvements in all phases of production—would result in costs below $10 per pound of uranium oxide (1975 dollars) during the first quarter of the next century.

Although the historical trend towards higher productivity in uranium production was reversed in the early 1970s, raising a legitimate question about achievement of the production cost cited above, this provides no basis for optimism about the future prospects for plutonium profitability. Among the most important factors responsible for the recent rise in costs of uranium production have been higher standards of worker safety and environmental protection. Because of the very much greater risks of radioactive contamination in the separation of plutonium, these have had a far greater impact on the costs of reprocessing. Since the early 1970s, reprocessing costs have increased, in constant dollars, by a factor of 10, while uranium costs have roughly doubled. Therefore, the prospect that a further heightening of concern in these areas might double uranium costs again should be no cause for hope among proponents of plutonium, since they would then, in all likelihood, be faced with reprocessing costs measured not in hundreds, but in thousands of dollars per kilogram. This would remove beyond question any lingering doubts about the diseconomies of plutonium fuels.

Perhaps natural uranium will become in the future far more expensive relative to plutonium than history and present evidence suggest is likely. Then, and not before, would be the time to decide whether the economic gains from use of plutonium were worth the increase in the risks of proliferation, environmental damage, and nuclear terrorism.

Organisation of Following Sections

These sections attempt to provide a clear picture of the role which reprocessing might play in the development of the nuclear industry. After an introductory explanation of the way in which nuclear power reactors convert uranium into electricity, the history and current status of the development of reprocessing in the United States and other countries are reviewed. There follow analyses of the various arguments for reprocessing, discussion of the economics of plutonium fuels, and reviews of recent, major analyses of the economics of reprocessing. The evidence for the claim that early introduction of plutonium fuels is necessary because uranium scarcity is upon us or looming just over the horizon is examined. Separate sections explain why support of the notion of uranium scarcity reflects inadequate understanding of the economics of the uranium market. Evidence upholds a high likelihood of obtaining enough low-cost uranium to support anticipated nuclear-electric capacity in the United States and the world until well into the twenty-first century.

An introduction to the nuclear fuel cycle: Nuclear reactors operate by sustaining a controlled " chain reaction ". In this chain reaction, when one atom fissions, it sends out a number of energetic neutrons which—on average—cause at least one more atom to break apart, which in turn sends out more neutrons which split another atom, and so on. When an atom splits, energy is released, and, in a power reactor, this energy heats water sufficiently to convert it to steam, which is then used to drive an electricity-generating turbine exactly as in a conventional, fossil-fuelled generating plant.

To create a controlled chain reaction, a reactor requires " fuel " containing atoms which are readily fissionable (capable of being split apart by relatively low-energy neutrons). The only such atoms which occur in nature are atoms of uranium-235 (uranium atoms with 235 neutrons and protons). These atoms comprise only about 0·7 per cent. of the atoms in natural uranium. Almost all of the remaining uranium atoms have 238 neutrons and protons—an " atomic weight " of 238—although trace amounts of other uranium isotopes—uranium atoms with different atomic weights—also occur in nature.

Uranium-238 is not readily fissionable and, thus, cannot directly serve as " fuel " for nuclear reactors. However, reactor fuel contains a high percentage of uranium-238, and the atomic reactions taking place in a reactor convert some of the uranium-238 atoms to readily fissionable isotopes of the element plutonium. These isotopes of plutonium contribute to the chain reaction and, thus, to the energy production in a nuclear reactor. Less plutonium is consumed than is created, however, so when the concentration of uranium-235 has fallen—and reaction by-products or " poisons " have risen—to the point where a chain reaction can no longer be sustained (when the nuclear fuel has become " spent "), there will be significant quantities of plutonium in the fuel.

The two classes of commercial power reactors: The isotopic composition of new and spent fuel depends upon reactor design. Two broad classes of power reactors are in commercial production at the present time. The most common class of reactor—and the only type produced at present in the United States—is the light water (LWR), in which the nuclear fuel is surrounded by normal water. Within this class, two different types exist: pressurised water reactors (PWRs) and boiling water reactors (BWRs).

Although these two types differ in design, their characteristics are quite similar from the standpoint of nuclear fuel. The second important class of power reactor is the heavy water reactor (HWR), in which the nuclear fuel is surrounded by deuterium oxide—commonly termed " heavy water " because it is chemically the same as water but contains deuterium, a naturally occurring isotope of hydrogen with an atomic weight of two. The Candu reactor produced by Canada dominates the commercial market for HWRs.

The composition and production of fresh reactor fuel: In light water reactors, readily fissionable atoms of uranium-235 make up 2 to 4 per cent. of the weight of uranium in new fuel, the remainder being uranium-238. Light water reactors cannot sustain a chain reaction with natural uranium, which contains only about 0·7 per cent. uranium-235. Thus, natural uranium must first be " enriched " in uranium-235 before it can serve as fuel for LWRs. Alternatively, addition of plutonium to natural uranium can substitute for enrichment in the production of LWR fuel. This will be considered later. Processes to enrich uranium are technically demanding and require substantial investments. Until recently, the United States was the only significant source of enrichment for reactor fuel, although a number of other countries are now constructing enrichment facilities for the civilian market.

In contrast to light water reactors, heavy water reactors operate with natural uranium. Heavy water absorbs far fewer neutrons than normal water; thus a chain reaction can be sustained with a very low percentage of readily fissionable atoms. Enrichment is not a necessary part of the fuel cycle of HWRs. Except for the enrichment process required for LWR fuel, production of LWR fuel and HWR fuel is very similar. Uranium of the appropriate isotopic composition is converted to uranium dioxide (UO_2), formed into cylindrical pellets, and inserted in metal tubes. These fuel " rods " are assembled into fuel " bundles ", in which form they are ready for insertion into a reactor.

Radioactive wastes and the composition of spent fuel: Because LWR fuels initially contain a higher percentage of fissile atoms, more fission reactions occur in LWR fuels than in HWR fuels before they are replaced. This " burnup " of fissile material is commonly measured by the heat produced per unit of fuel. Typical LWR fuels have burnups over the range of 12,000 to 33,000 megawatt-days-thermal per metric ton of uranium metal (MWD_{th}/MTU), with burnup nearer the upper figure appropriate for replacement fuel loadings. The burnup of fuel in Candu reactors, the dominant brand of commercial HWR, is about 7,500 MWD_{th}/MTU. Generally speaking, the radioactivity, heat output, and residual fissile materials of spent fuel all increase with fuel burnup.

Uranium-235 and fissile plutonium are potentially useful for reactor fuel, if recovered from the spent fuel. In addition, spent fuel contains many other radioactive isotopes which are created by the splitting of atoms (fission products) and by the capture of neutrons by uranium without fissioning. These " transuranics "—including plutonium—and the fission products are extremely dangerous to life processes. In sufficiently high concentrations they can cause death in a short time. In lower concentrations, they can lead to increased incidence of cancers and birth defects. An important requirement of spent fuel management is to provide great assurance that these dangerous, radioactive substances will be kept isolated from the general environment. This is no simple task. The radioactivity of spent fuel is intense, and heavy shielding is required to protect the workers who handle it. Some of the isotopes are extremely long lived and must be kept isolated for hundreds of thousands of years. The radioactivity generates heat, and, especially during the early years after discharge, adequate cooling of the spent fuel is an important requirement. Further, the radioactivity will cause physical-chemical changes in any materials used to contain the radioactive wastes, creating uncertainty about the long-term structural integrity and resistance to leaching by water of any containment material.

Reprocessing, " recycle " and the plutonium fuel " cycle ": " Reprocessing " is the term commonly used to refer to the mechanical-chemical process used to separate potentially useful elements—mainly uranium and plutonium—from other components of spent fuel. Use of the recovered products in fresh fuel is generally called " fuel recycle ". Use of the term " recycle " is reasonably appropriate when applied to use of recovered uranium-235, but it is misleading when applied to plutonium. Plutonium " recycle " implies that plutonium fuel has already been used once and is merely to be refabricated and reused. But, without reprocessing, no plutonium will exist separately or in fresh reactor fuel. What is commonly termed implementation of plutonium " recycle " is in fact introduction of a plutonium fuel cycle and ought to be labelled as such. Also, plutonium fuel is commonly referred to as mixed-oxide (abbreviated as MOX) fuel because it contains a mixture of uranium and plutonium oxides. Again, this obscures the meaning to the non-expert. Plutonium fuel is an accurate term and ought to be used instead.

The term " nuclear fuel cycle " is so firmly embedded in the language that there is little hope of altering it. But, everyone should be clear about the reality which underlies the term, so that it does not evoke a false image. The term refers to the totality of activities involved in producing nuclear fuels, using them in reactors, dealing with the spent fuel, and managing all of the radioactive products released or produced by these activities. It is a " cycle " only in the sense of including all aspects of the fuelling of nuclear reactors, from the start of the process to the final disposition of the radioactive wastes. It is in no sense a circular, closed loop. The uranium fissioned in a reactor is transmuted into other elements. These other elements never return to their original form so that they can be " cycled " again in a reactor, although some of the elements, such as plutonium, can be further transmuted in a reactor. The new elements produced by the fission reactor are, in fact, the vast bulk of the radioactive " wastes " which must be dealt with in the nuclear fuel cycle.

A common appeal by industry today is for the United States government to take the actions necessary " to close the nuclear fuel cycle ", by which it is generally meant that it should give approval for reprocessing and for use of plutonium fuel. What could be more natural and desirable than " closing " a cycle? Used in this way, the term "fuel cycle " evokes an image which, incorrectly, supports reprocessing as a normal, logical aspect of fuel management. Further, to talk about a " closed fuel cycle " suggests that there will be no problem of radioactive wastes, that these wastes will somehow be consumed or transformed within this magical " cycle ". But, of course, this is erroneous. Whether or not spent fuel is reprocessed, most of the radioactive products of the fission reactions will remain and will need to be securely isolated from the environment.

At present, all spent fuel in the United States is placed in temporary storage in water-filled pools. As the radioactive products in spent fuel must be securely isolated for hundreds of thousands of years, temporary storage is hardly a satisfactory terminus of the fuel " cycle "—although, until a few years ago, few nuclear planners, manufacturers, or users were particularly concerned about the lack of any agreed means for ultimate storage of radioactive wastes. Thus, the present fuel cycle can be accurately termed unsatisfactory, but to describe it as " not closed " is misleading.

The term " throwaway fuel cycle " is often used by advocates of plutonium fuel to refer to the alternative of placing spent fuel directly in geological storage, without reprocessing. Then, the issue of whether to permit plutonium fuels is posed as a choice between a " closed fuel cycle " and a " throwaway fuel cycle ", leaving little doubt in anyone's mind about which is more desirable.[18] More accurate terms for the latter would be " direct disposal of spent fuel " or " direct permanent storage of spent fuel ".

The History and Current Status of Reprocessing

Reprocessing of uranium first occurred in the Manhattan Project. One of the bombs dropped on Japan (" Fat Boy ") used plutonium obtained by reprocessing uranium irradiated in reactors at Hanford, Washington. Production of plutonium in special reactors and separation by reprocessing has been a continuing component of every nuclear weapon programme in the world.

Given the early experience with reprocessing, it is not surprising that the use of the recovered plutonium and uranium in new fuel was assumed to form an intrinsic part of the civilian nuclear power programme. Also, early projections of uranium resources showed that standard thermal reactors would consume all this uranium before they could make a significant contribution to energy production. The proposed solution has always been the development of " breeder reactors ". Such reactors

---

[18] See, for example, Cholister, R. J., *et al.*, *Nuclear Fuel Cycle Closure Alternatives* (Barnwell, S.C.: Allied-General Nuclear Services, April 1976), p. 1; and U.S. Energy Research and Development Administration, *Benefit Analysis of Reprocessing and Recycling Light Water Reactor Fuel*, ERDA 76-121 (Washington, D.C.: ERDA, December 1976) pp. 1–2.

" breed " more fissile material than they consume—by converting uranium-238 to plutonium, or by converting thorium-232 to uranium-233, another readily-fissionable material—thus greatly extending the nuclear fuel supply. Plutonium breeders have been the favoured type of breeder in the United States. Reprocessing and plutonium fuel are indispensable for plutonium breeders; thus, there was never any question about the necessity for reprocessing in the minds of nuclear planners in the United States.

In the early years, reprocessing technology was highly classified because of its applicability to the production of nuclear weapons. But the " atoms for peace " programme, initiated in 1954, and interest by the Atomic Energy Commission in promoting commercial reprocessing led to almost total declassification of this technology in 1955 and 1956.[19] The final details of the chemistry of the standard Purex process were declassified in 1958. The AEC was instrumental in establishing the first commercial reprocessing plant in the United States at West Valley, New York. Reprocessing in Europe was also encouraged through further declassifications for international conferences and through direct technical assistance. These activities gave great impetus to international development of a commercial reprocessing industry.

In 1970, the reprocessing industry appeared to be developing in an orderly way.[20] The first commercial plant in the United States was in operation, the General Electric Company was building a plant, and a larger one was in final planning stages. Pilot plants for reprocessing commercial fuel had started operating in Belgium (1966), Italy (1969), and Germany (1970). There was also excess capacity at the Magnox (uranium-metal) fuel reprocessing plants at Windscale in the United Kingdom and La Hague in France. By adding new " head-end " facilities, these plants could be used to reprocess the uranium-oxide fuels used in commercial light water reactors. Thus, in 1970, there seemed little reason to doubt that there would be use of plutonium fuel.

Developments since 1970, however, have raised serious questions about the desirability of proceeding on this course. Many problems have been encountered in technology, regulation, and economics, and also at the political level.

The existence of technical problems was revealed by the less than satisfactory performance of the early oxide reprocessing plants. The plant of Nuclear Fuel Services (NFS) at West Valley, New York, encountered many difficulties in its six-year operating history and was shut down in 1972 for extensive modification and expansion. After processing only 100 tonnes of oxide fuel in three years, the Windscale plant suffered a small explosion in the oxide " head-end " facility in 1973. The investigation which followed revealed unexpected process deficiencies and the

19 Information on declassification and early efforts by the AEC to encourage reprocessing is from Clapp, P., " The Declassification of the Purex Process ", Arms Control and Disarmament Agency, unpublished, undated.
20 The following discussion draws heavily on Rippon, Simon, " Reprocessing—What Went Wrong? ", *Nuclear Engineering International*, XXI, 239 (February 1976), pp. 21–27.

decision was made to redesign extensively the head-end. The General Electric Company's plant at Morris, Illinois, which was based on an innovative technology, never operated, and has apparently been written off as a total failure.

The technical problems stem largely from the high radioactivity of oxide fuels, which is 10 to 25 times higher than the uranium metal reprocessed in the weapons programmes. None of these problems appears to be insoluble, but they create uncertainties about costs and reliability of the coming generation of plants.

Early reprocessing plants were built when the nuclear industry was relatively small and uncontroversial. By the early 1970s, however, there began a process of upgrading of safety standards that is continuing today. In the United States, the general policy was adopted of setting permissible radioactive emission levels " as low as practicable ", and this has recently been modified to a more stringent standard to be " as low as reasonably achievable ". Because of these criteria, advances in technology have led to increasingly stringent requirements.

The reprocessing industry has been especially affected by the growing concern with radioactive wastes. For example, the early plant at West Valley, New York, simply placed its highly radioactive, liquid wastes in a large storage tank, with no provision for converting them to a form suitable for long-term isolation—approximately one million years—from the environment. Recently instituted regulations require any future reprocessors to solidify high-level wastes within five years of the time they are produced. Other proposed regulations would increase the fraction of waste products which would need to be handled as high-level wastes. Many important requirements are still unspecified.

Another concern is the potential for " nuclear terrorism ". Plutonium, because it can be fabricated into a nuclear weapon, has a widely publicised terrorist potential. It is also an extremely toxic, radioactive poison. A threat to disperse plutonium-oxide powder widely over a major city would need to be treated with the greatest seriousness, and if carried out, would cause great disruption of normal activity, require significant evacuation, and necessitate an enormous decontamination programme. New and much stricter standards of physical security for plutonium and plutonium fuels are still evolving.

Changing regulations and uncertainty about future requirements have created substantial problems for the commercial reprocessing industry in the United States. Nuclear Fuel Services, Inc., owners of the facility at West Valley, New York, which shut down for modification and expansion in 1972, has announced that it will not attempt to reopen the plant in view of economic and regulatory difficulties. It estimated that preparing the plant for operation would require $600 million and a dozen years—as compared with its original estimate in 1972 of $15 million and two years—with no assurance that a licence would be issued.[21] A major new American plant—that of Allied-General Nuclear Services

---

[21] " Getty Oil Subsidiary Says It Won't Reopen Nuclear Fuels Facility ", *Wall Street Journal*, 23 September, 1976, p. 20.

(AGNS) at Barnwell, South Carolina—was another victim. Although the basic reprocessing plant was largely completed by 1975 at a cost of $250 million, the operating licence cannot be obtained until after the decision by the Nuclear Regulatory Commission on the use of plutonium fuel. Furthermore, the Commission has ordered that additional facilities be built before the plant can be licensed. Allied-General Nuclear Services have halted all construction and design work. Current estimates are that completing the additional facilities would bring the total cost to over $1 billion.

Regulatory concern has had far less impact abroad, in part because some countries have not established independent regulatory authorities, and in part because reprocessing and recycle has proceeded at a slower pace.

The United Kingdom has recently placed reprocessing under control of the independent Nuclear Installations Inspectorate (NII), which does not appear to have created any special problems for the government organisation (BNFL) with responsibility for reprocessing. Plans to build a new reprocessing facility at Windscale have, however, been delayed by the decision of the Secretary of State for the Environment to hold a public inquiry before issuing a construction permit.[22]

The nuclear industry in France has not been subjected to independent regulation. However, a recent news item reported that "The French government grouped CEA's nuclear safety and radiation protection services under a new body . . . which will be funded by the Ministry of Industry rather than by the CEA."[23] Other details of this news item, however, make it clear that the degree of independence of the safety board from CEA will be limited.

Germany has only a pilot reprocessing plant in operation, and current planning does not envision that a large-scale plant be built until after 1985. But the pressure of the regulatory authorities in Germany, and in Japan, has been almost opposite in direction to that in the United States: because radioactive waste is such a major concern, regulatory officials have exerted great pressure to develop acceptable solutions for the handling of spent fuel, and the solutions most commonly being considered involve reprocessing to permit the radioactive wastes to be incorporated into glassified solids. Set against this pressure to move forward on reprocessing is significant local opposition to the siting of any nuclear plants, and opposition to reprocessing can be expected to be especially intense.[24] In Japan, public opposition to reprocessing is probably a more important influence than the pressure of regulations. Approval has been received for operation of a large pilot plant (210 tonnes per year), which had been delayed

22 *Nucleonics Week*, XVIII, 2 (13 January, 1977), p. 8.
23 *Nucleonics Week*, XVII, 46 (11 November, 1976), pp. 7–8.
24 *Nucleonics Week*, XVII, 40 (30 September, 1976), p. 12, reports that " Environmental protests of unprecedented proportions threaten the German waste management centers planned for full load operations in 1988–89. . . . Antinuclear groups in West Germany have decided that by spiking the waste management center they can kill the whole nuclear program. Observers believe that the occupation at the site of the proposed Wyhl nuclear plant will be dwarfed by the storm now building up over the waste management center."

in starting production until at least 1977.[25] Public opposition to the building of a full-scale reprocessing plant is intense, and it is unlikely that
approval can be gained in the current environment of opposition to nuclear
development in Japan. India has no independent nuclear regulatory body,
and no opposition can be expected to plans of the Indian AEC to expand
reprocessing capacity by building two plants in addition to the one already
in operation.

Technical difficulties, regulatory changes, and escalating costs of construction have combined to increase reprocessing costs at an astounding
rate. The capital costs of the West Valley, New York, plant of 1 tonne per
day capacity were $35M (1963–67 dollars). The Barnwell, South Carolina,
5 tonne per day facility of AGNS will probably cost over $1·0 billion (1976
dollars) before—and if—it opens. Costs per kilogram of fuel reprocessed
have grown apace. A few years ago, reprocessing costs were estimated at
about $30 per kilogram.[26] Including temporary storage and treatment—
but not final disposal—of wastes, costs are now estimated at up to $300
(1976 dollars) per kilogram. After allowing for inflation, this means a rise
by nearly a factor of seven in two years. Nor is there any assurance that
the escalation of costs is over. There has also been a rapid rise in the
estimated cost of fabricating plutonium fuel. From 1971 to the end of
1974, the inflation-adjusted fabrication premium [27] quoted by fabricators
in the United States for plutonium fuel rose from about $1.50 to $6.50
(in 1975 dollars) per gramme of fissile plutonium.[28] This is equivalent to
an increase of about $30 per kilogram in reprocessing costs. These in-
creases have more than offset the recent substantial rises in costs of
enrichment and uranium ore, rises which favour profitability of
reprocessing.

Not only have reprocessing plants in the United States been closed or
have had work on them stopped, but there is evidence of growing realisation in Europe that reprocessing will prove to be unprofitable. In December
1975, Germany, England, and France agreed to cooperative, non-competi-
tive development of reprocessing capacity. In a detailed agreement approved
by the Commission of the European Communities, these three countries
formed a joint company, United Reprocessors GmbH (URG), and agreed
upon market shares and the schedule of additions to capacity in each
country.[29] Given the strong competition among these countries in other
areas of nuclear power, this agreement strongly suggests that they do not
view reprocessing as a likely source of profit. Additional confirmation of
the European position is provided by a recent report of United Reprocessors' talks with European utilities:

[25] *Nucleonics Week*, XVII, 27 (1 July, 1976), p. 9. Production has now started.

[26] See U.S. Atomic Energy Authority, *General Environmental Statement on Mixed
Oxides (GEMSO)* (draft), WASH-1327 (Washington, D.C.: U.S. Government Printing
Office, August 1974), Vol. IV, pp. ix–41.

[27] The extra charge above that made for fabricating uranium oxide fuel.

[28] Stoller, S., *et al.*, " Report on Reprocessing and Recycle of Plutonium and Uranium—
Task VII of the EEI Nuclear Fuels Supply Study Program " (Edison Electric Institute, 30
December, 1975), Appendix VI, pp. 86–88.

[29] Commission Decision of 23 December, 1975, Relating to a Proceeding under Article 85
of the EEC Treaty (IV/26.940/a—United Reprocessors GmbH), reported in *Official Journal
of the European Communities* (26 February, 1976), pp. No. L 51/7—L 51/14.

. . . URG is seeking firm reprocessing contracts combined with advance payments from utilities to help finance the heavy investment costs, and the pressing
need for more reprocessing capacity in Europe puts URG into a strong bargaining position. . . . According to an URG executive, utilities had " in
principle accepted the reasoning that reprocessing will in the future not be
possible on a purely commercial basis " but will require a cooperative effort.
A German utility source says future reprocessing costs are expected to amount
to $350–400/kg. . . . Of course, part of the reprocessing cost will be offset by
the value of the reprocessed uranium/plutonium. But the fact remains that
reprocessing, which up to 3–4 years ago was estimated to cost less than the
retrieved fuel, now stands to become a rather substantial addition to the
overall nuclear kwh cost.[30]

Is Plutonium Really Necessary?

The nuclear community commonly holds that plutonium fuel is essential
to nuclear power, that there is no reasonable alternative to it in the long
run, and that the sooner we begin using it, the better off we will be. However, first, there is no technical necessity to use plutonium as nuclear
fuel so long as supplies of uranium-235 are available. There is enough
uranium-235 in the world to fuel all present and planned reactors for
decades and even for centuries, although some may be very expensive to
extract. Plutonium use in the current generation of nuclear reactors is,
therefore, a question of economics and not of technical necessity. Second,
statements about the desirability of early plutonium recycle generally
refer to light water reactors. Heavy water reactors use natural uranium as
fuel and provide little economic incentive for using plutonium fuels.
Plutonium recycle for HWRs would still be very unprofitable under conditions which might make it profitable—in the narrow economic sense—
to recycle plutonium for LWRs. Third, the assertion that use of plutonium
fuel will be necessary in the long run implies that we must eventually rely
on " breeder " reactors to satisfy our needs for electrical power. But, in
discussing the " necessity " for breeders, we are once again in the realm
of economics. Breeders will never be essential in the strictly technical sense,
but rather may provide an economically attractive alternative source of
power. It is very unclear, when, and even if, breeders will become economically superior to other alternatives, but it will certainly not be before the
1990s.

The conservation argument: The potential saving of uranium is often cited
as an important reason for the use of plutonium fuel, entirely aside from
the economics. For example: " With this dramatic uranium conservation
benefit to be achieved through reprocessing, it is simply unacceptable to
continue to lend credence to the ' throwaway ' fuel cycle." [31]

[30] McQueen, Silke, " United Reprocessors GmbH (URG) in Talks with European
Utilities ", *Nucleonics Week*, XVII, 29 (22 July, 1976), pp. 7–8.
   Less than a year later, the European outlook for the profitability of reprocessing had
worsened dramatically: URG was reported to be demanding 10-year firm reprocessing contracts, at $800 per kilogram, with customers financing the entire planning, construction, and
operation for the 10 years (but without acquiring any equity interest), with return of radioactive wastes to the customers, and no guarantees that reprocessing would actually be
accomplished; *ibid.*, XVIII, 14 (7 April, 1977), p. 7.
[31] Price, W. J., " Reprocessing Incentives " presented at the AIF Fuel Cycle Conference,

A more restrained statement is the one in a recent ERDA study:

Although favourable fuel cycle economics demonstrate an advantage for recycling spent LWR fuel, uranium availability and resource conservation are also important considerations. Without recycle the residual fissile content of spent fuel could never be utilised. . . . Recycle of uranium and plutonium would result in the generation of additional energy from a given amount of natural uranium and thus assist in achieving energy independence.[32]

Thus, at the least, uranium savings are cited as a " bonus " from " recycle " and, at the most, as sufficient reason to undertake reprocessing. These arguments are seriously deficient both logically and practically. They violate the most elementary principles of economics. They count " savings " but ignore costs. By similar reasoning, avoidance of reprocessing would " save " over $50 billion in the next 25 years, according to the assumptions in the base-case analysis of the ERDA study. The savings would be achieved by conserving the resources which otherwise would be devoted to reprocessing ($45.1 billion) and plutonium-fuel fabrication ($5.1 billion). The possibility of saving $50 billion certainly provides a sufficient reason to forgo reprocessing.

Of course, the proponents of recycling will quickly point out that the United States will have to spend more on uranium mining and enrichment if it forgoes recycling. But the converse is also true—that the United States must expend a great many resources on reprocessing, fabrication of plutonium fuels and radioactive waste management to reduce the consumption of uranium ore by reprocessing and use of plutonium fuel. In deciding whether or not reprocessing is economically beneficial, the additional resources spent must be balanced against the savings in resources. To ignore the expenditures of resources is the most naïve type of analytical error.

In this age of " limits " to growth, conservation of resources has become fashionable. The conservation of uranium, a limited energy resource, may seem to be more important than conserving the resources which would need to be expended to permit use of recovered uranium and plutonium. But how much of these other resources, including natural resources, ought we be willing to expend to save a given quantity of uranium? How do we value these other resources relative to uranium? Is a pound of uranium worth a ton of steel, a thousand feet of copper cable, a thousand gallons of nitric acid, a cubic foot of lead, or a thousand man-hours of labour? Is a pound of uranium more irreplaceable than the hours of individual lives which would be expended in constructing reprocessing facilities?

The generally accepted answer is that, in choosing among alternative economic activities, one should weight the various resources involved by their market prices and then choose the activity which costs the least. In the absence of market imperfections, this procedure results in the highest value of production from a given set of resources. It is the prescription for

1976, Phoenix, Arizona, 23 March, 1976. Mr. Price is the executive vice-president of Allied General Nuclear Services, which is the owner of the reprocessing plant at Barnwell, South Carolina.

[32] *Benefit Analysis of Reprocessing and Recycling Light Water Reactor Fuel*, ERDA 76–121 (Vienna : ERDA, December 1976), p. 3. Referred to hereafter in this section as " the ERDA study ".

efficient production on which our free enterprise economy is based. But, of course, weighting by market prices is exactly the procedure used in the economic analyses of reprocessing.[33] Our economic analysis will indicate that the resource-conservation " benefits " of reprocessing would probably be negative and, in any event, insignificant relative to the total resources required to produce and deliver nuclear power.

But what about contentions that the uranium saved by recycling would permit the United States to fuel a substantially greater number of nuclear reactors with its limited uranium supply? For example, the recent ERDA study states:

The reduction in uranium ore requirements with reprocessing is equivalent to providing the additional fuel necessary to increase the installed electrical generating capacity of 500,000 MWe in the year 2000 assumed in the base case of this report to a level of 650,000 MWe.[34]

However, the supply of uranium at sufficiently high prices is practically unlimited relative to short-term demands. Including the uranium in sea-water, the United States has access to 6,000 times the potential uranium savings for 1975–2000 estimated in a recent ERDA study of reprocessing. The number of nuclear reactors installed, if politically acceptable, will depend upon the demand for electricity and the relative cost of electricity generated by nuclear power, neither of which would be significantly affected by implementation of reprocessing. Recycling is not the only means of conserving uranium resources. An alternative is to substitute enrichment for uranium in LWRs. By revising operating parameters to utilise 70 per cent. of the uranium-235 in feed—a change which would lower total fuel costs at present prices—ERDA could reduce uranium requirements by 33 per cent. relative to those projected under some proposed operating plans. Savings could also be achieved by shifting emphasis from light-water-reactor technology to heavy-water-reactor technology of the Candu type. If fuel enriched to 1 per cent. uranim-235 were used in Candu reactors, their uranium consumption would be only 50–60 per cent. of that of LWRs. These savings are greater than those realistically attainable through the use of plutonium fuels.

The " waste management " argument: Reprocessing has been promoted, especially in Europe and Japan but also in the United States, as a means of simplifying the management of the radioactive poisons contained in spent fuel. There is an urgent need to demonstrate that spent fuel can be safely managed. But reprocessing will greatly complicate this demonstration as compared to direct placement of spent fuel rods into geologic containment. First, the original radioactive wastes in the spent fuel would contaminate much larger volumes of process chemicals and materials in the

[33] If proponents of recycle wish to argue that the uranium saved by recycle is " worth " more than its market value—because it is a natural resource—they also must accept that all uranium is worth more than its market value. (There is no reason that " saved " uranium should be valued more highly than consumed uranium.) The policy prescription which follows from this belief is a tax on the consumption of all uranium, to raise the market price to its " true worth ". Then, the market will act to conserve uranium in the most efficient way.

[34] Benefit Analysis of Reprocessing and Recycling Light Water Reactor Fuel . . . p. 3.

reprocessing facility. Development of satisfactory methods of fully oxidising combustibles and of compacting non-combustibles may be necessary before contaminated solids can be accepted for geologic isolation. The most dangerous radioactive materials (high-level wastes) would be converted during reprocessing from a very stable, solid ceramic to acid-solution form and then would need to be reconverted to acceptable solid form.[35] Such conversion processes are difficult, will create more wastes, pose additional regulatory problems, and have not yet been demonstrated on a commercial scale. Further, most of the solids being considered for reprocessing wastes appear to be less desirable forms of waste than the original spent-fuel pellets, which are highly stable ceramics designed to withstand the extreme operating temperatures and pressures of nuclear reactors.

Second, the reprocessing facility itself would become contaminated, thus additionally complicating the demonstration that management of spent fuel will not lead to the radioactive contamination of the environment. Third, reprocessing would release radioactive gases which would otherwise be contained within the spent fuel rods. The technology for trapping and retaining some of these gases, especially tritium, needs extensive development before it can be considered " demonstrated technology ". Fourth, if plutonium were separated, possibilities would be created for illicit diversion of plutonium and nuclear terrorism. Even though stringent safeguards and physical security measures may be imposed, the mere existence of these possibilities seems likely to create doubts in the public mind about the safety of spent fuel management.

Reprocessing as a part of a long-term waste-storage programme: Spent fuel contains radioactive isotopes which will persist for hundreds of thousands of years, making environmentally safe storage of these wastes a matter of great controversy. The possibility exists, however, of removing essentially all of the long-lived transuranic isotopes by reprocessing, leaving a residual waste product which decays to safe levels in less than 1,000 years. Separated transuranics could be recycled into reactors and transformed into shorter-lived isotopes. However, the reprocessing system required for this scheme bears no relation to the commercial reprocessing facilities currently being considered, since these are designed to leave all transuranics except plutonium in the waste, and even the percentage of plutonium removal would not be sufficient to meet the standards required for the advanced waste-storage programme. Nonetheless, as a solution to the waste problem, even very high reprocessing costs might be tolerated: at $500 per kilogram reprocessing would add only 2 mills per kwh to the cost of electricity.

Such a system is in the preliminary stages of conceptual design in the United States. Development to a practical stage of application, if it occurs at all, is many years in the future. No other countries, to our knowledge, are seriously contemplating such a waste handling system.

Small-scale reprocessing facilities: Small-scale, research and development reprocessing facilities present a special problem for policy. For example,

---

[35] The borosilicate glass most commonly envisioned as the solid form for reprocessing wastes is less resistant to leaching and thermally less stable than spent fuel.

a government could quite justifiably argue that it needs to gain experience in reprocessing to prepare for the eventuality of an economic breeder. How should one judge the desirability of " pilot or small-scale reprocessing facilities " ? Again the question is whether the economic benefits justify the military risks. The costs, benefits, and risks all diminish with smaller size, but how the balance among these alters is not immediately obvious. We suggest that the most important considerations in assessing small-scale facilities are the production rates and agreed-upon stockpiles of plutonium. The military risks are dominant, since they far outweigh any possible economic gains from small-scale facilities. On the other hand, it should be possible to accomplish any desired research and development involving reprocessing without production or accumulation of dangerous amounts of plutonium. A facility which can produce only a few hundred grams of plutonium per year is of relatively little concern, if agreement can also be obtained to limit total plutonium inventories to about a kilogram. A reasonable objective for policy should be to gain agreement that plutonium production rates and stockpiles should be as small as possible relative to the amount needed for a nuclear weapon.

Comparative Costs of Uranium and Plutonium Fuels

Comparing the cost of a plutonium fuel cycle with that of a natural uranium fuel cycle involves many technical and economic complexities, but the major conclusions can be understood from a simplified comparison of the fuel cycles. Figure 2 presents such a comparison. It considers alternative dispositions of mature-cycle spent fuel from pressurised water reactors, the type of fuel which it is most profitable to reprocess. The various processes involved are indicated by rectangular boxes, with the estimated costs noted within, and the material flows by rectangles with clipped corners, with the double-lined ones giving the total estimated costs. The question marks indicate processes or materials the costs or values of which have not been specified in the comparison. Total costs exclude consideration of the costs of final storage of spent fuel or plutonium-fuel-cycle wastes and the costs of safeguards and physical security for plutonium fuel. These costs are omitted because existing studies are inadequate to provide meaningful estimates of their magnitudes and our own analyses of costs are not yet completed. However, as shown in a later section, consideration of volumes and heat-generation characteristics of radioactive wastes suggests that inclusion of final storage costs will add more to the plutonium fuel cycle than the uranium fuel cycle.

The uranium fuel cycle: In the uranium fuel cycle, production of 1 kg of enriched fuel requires that 6 kg of natural uranium containing 44 gm of uranium-235 be mined and milled, and then converted to gaseous form in preparation for the enrichment process, in which five separate work units are applied to the 6 kg of natural uranium. Of the uranium-235 entering the enrichment process, approximately 10 gm finish in the waste stream—mixed into approximately 5 kg of uranium-238—and 34 gm in the 1 kg of enriched product. The enriched uranium is converted to oxide form and fabricated into fuel rods. The fuel is then irradiated in a power

FIGURE 2

The Uranium and Plutonium Fuel Cycles

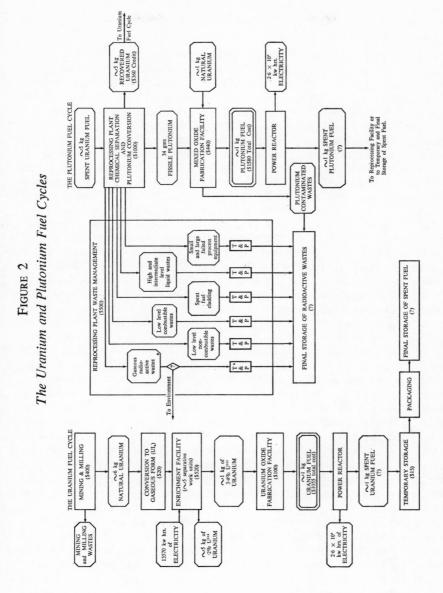

reactor. When fully irradiated, the spent fuel is removed and placed in
temporary storage, then transported to a central facility for packaging
and final storage. The costs of managing the fuel after discharge from
the reactor must be added to the cost of preparing the new fuel in order
to obtain the total fuel cycle cost (only costs of temporary storage are
included in the total of $1,055 per kg shown in Figure 2). The important

items in this total are the cost of uranium (assumed to be $70 per kg, including conversion to gaseous form) and the cost of enrichment (assumed to equal $100 per separative work unit).

The plutonium fuel cycle: The plutonium fuel cycle begins with 5 kgs of spent uranium fuel. The fuel is transported to the reprocessing plant, where the spent fuel rods are chopped into pieces and the heavy metals it contains (uranium, plutonium, other transuranic isotopes) and fission products are dissolved in acid solutions. The uranium and plutonium are then precipitated from the acid solutions, leaving the fission products and other metallic radioactive isotopes in the solutions. These solutions then contain most of the radioactive wastes originally contained in the spent fuel. Figure 2 indicates the many types of waste and treatment processes involved in reprocessing. There is a widespread belief that reprocessing will simplify the management of the radioactive wastes in spent fuel.[36] As the figures suggest, nothing could be further from the truth.

The costs of the reprocessing functions, which include waste management and conversion of recovered plutonium to oxide form, are highly uncertain. The cost used in Figure 2 is $300 per kilogram. Two thirds of this cost are assigned in Figure 2 to the chemical separation and plutonium conversion functions and the remainder to the waste management functions. These proportions are only approximate. For the type of spent fuel assumed in Figure 2—reload fuel from PWRs—approximately 5 kg of uranium and 34 gm of plutonium are recovered from the 5 kg of spent fuel. Each kg of recovered uranium is approximately equivalent to 1·1 kg of natural uranium.[37] The recovered uranium goes to the uranium fuel cycle, providing a credit which partially offsets the cost of reprocessing.

Because of the toxicity and radioactivity of reactor-grade plutonium, plutonium fuel must be fabricated in a special plant using automated or shielded, remote-handling equipment. Commercial-scale processing equipment to fabricate reactor-grade plutonium—which is much more radioactive than the weapon-grade material used in development work so far—is still in the development stage; thus the costs of this operation are very uncertain. Quite clearly, though, the cost will be much higher than for fabrication of uranium fuel. The cost estimates used in Figure 2 are $370 per kg for plutonium fuel fabrication[38] and $100 per kg for uranium fuel fabrication.

The total cost of preparing plutonium fuel for the items specified in Figure 2 is $1,580 per kilogram. Safeguards and physical security costs have been omitted. Inclusion would raise the cost. Including the costs of final storage of radioactive wastes could, in theory, either lower or raise the cost. Manufacturing plutonium fuel avoids the costs associated with placing 5 kg of spent fuel (per kg of plutonium fuel) into final storage, but incurs the costs associated with final storage of the wastes generated by reprocessing 5 kg of spent fuel and by fabricating 1 kg of plutonium fuel. Depend-

---

[36] See, for example, " Scaling up Windscale ", *The Economist*, 18 December 1976, p. 15.

[37] The recovered uranium is contaminated with undesirable isotopes, which degrades its value and creates problems in use.

[38] The cost shown in Figure 2 for plutonium fuel fabrication includes $70 for the uranium required

ing upon which costs are the greater, the fuel cycle cost of plutonium fuel could be raised or lowered when such storage is taken into account. As shown later, consideration of both heat generation characteristics and volumes of the alternative waste forms, *i.e.*, spent-fuel rods versus reprocessing and fabrication wastes, creates a presumption that direct placement of spent fuel in final storage will be cheaper.

The final item in the plutonium fuel cycle is the cost—or credit—for managing the spent plutonium fuel. To our knowledge, no detailed analyses are available of the effects of these factors on the cost of managing plutonium fuel. Preliminary analysis indicates that the value of spent plutonium fuel would be quite low—and possibly negative.

Recent Evaluations of the Economics of Reprocessing

Uncertainties in costs of the components of the fuel cycle combined with the sensitivity of the economics of reprocessing to some of the prices involved, have permitted different investigators to report widely differing estimates of the profitability of reprocessing. Most analyses have been published by those with a predisposition towards reprocessing, and the values assumed for key items—especially reprocessing charges, uranium prices, and relative disposal charges for reprocessing wastes and spent fuel—have been such as to show a net gain from reprocessing.

FIGURE 3

The Relative Costs of Uranium and Plutonium Fuels in
Recent Evaluations of Reprocessing
(Constant Dollars per Kilogram of Heavy Metal)

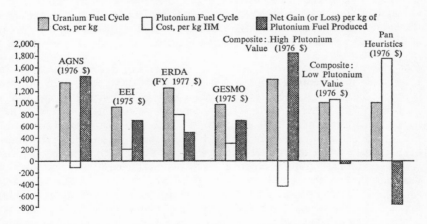

NOTE:
    Estimates omit costs of final storage of spent fuel and reprocessing waste.
    AGNS is Allied General Nuclear Services, *EEI* is Edison Electro Industry, *GESMO* is the *General Environmental Statement on Mixed Oxides* by the Nuclear Regulatory Commission, *ERDA* is the Energy Research and Development Association.
    SOURCE: Taylor, V., *The Economics of Plutonium and Uranium* (Los Angeles: Pan Heuristics, 30 April, 1977).

Several important findings emerge from comparing the results of five major studies (Figure 3). Most striking is the extremely wide variation in results. For the four studies conducted outside Pan Heuristics, the net gain per kilogram of plutonium fuel manufactured ranges from a "high" of $1,455 to a "low" of $499. In percentage terms, the range is from 108 per cent. to 41 per cent. of the cost of uranium fuel. Considering that all these studies were published within eight months of one another, this is a remarkable range. The actual range of variation in profitability ·implied by the four studies is even greater than the range of their results. The composite high and low plutonium values reported in Figure 3 were derived by combining the estimates of fuel cycle items from these four studies (all converted to 1976 dollars). On this basis, the net value per kilogram of plutonium fuel ranges from a gain of $1,840 to a loss of $40.

The similar results of the analyses by ERDA and the *General Environmental Statement on Mixed Oxides (GESMO)* mask very important differences in assumptions about the cost of fuel cycle components. ERDA uses a relatively high estimate for reprocessing cost ($280 per kg, including the cost of spent fuel transport), but is able to show a significant positive gain from reprocessing because it uses a very high price for uranium ($43 per pound of uranium oxide). *GESMO* uses a much lower price for uranium ($30.50) but also shows a significant net gain because it uses a low estimate for reprocessing costs. The results of the four studies depend to an important extent on the assumption that final storage of spent fuel will cost substantially more than final storage of radioactive wastes from reprocessing and plutonium fabrication. In the four studies, the savings in final storage costs per kg of plutonium fuel produced range from $240 to $360. In the studies by ERDA, *GESMO* and Edison Electric Institute (EEI), they account for about one half of the total net gain calculated. The empirical basis for estimating these savings is extremely deficient, and theoretical considerations suggest reprocessing will create additional disposal costs rather than savings.

Pan Heuristics plutonium fuel cycle cost estimates: Estimates by Pan Heuristics of the components of fuel cycle costs differ far less from those of the other four studies reviewed than might be inferred from a comparison of the final results. The uranium cost in this analysis is only moderately lower than that used in the EEI and *GESMO* studies—$70 per kg versus about $87 per kg (in 1976 dollars). The enrichment charge of $108 per kg is moderately higher than $81 per kg used in *GESMO* (in 1976 dollars) a difference which favours the case for reprocessing. The plutonium fabrication premium of $8.64 per gm of fissile plutonium (in 1976 dollars) is identical to that used in the EEI study. (The uranium fabrication cost cancels out in the net gain calculations). Only in the case of reprocessing costs is the Pan Heuristics estimate more than moderately higher than that used in another analysis. The reprocessing cost in the Pan Heuristics analysis exceeds the ERDA estimate, the next highest, by $79.00 (in 1976 dollars), allowing $20 per kg in the ERDA estimate for the transportation of spent fuel. But the high estimate of this item in the fuel cycle does not account for the Pan Heuristics estimate that reprocessing will prove unprofitable. If the ERDA reprocessing charge and the *GESMO* uranium and enrichment costs were substituted for the values used in the Pan Heuristics analysis, reprocessing would still show a loss of about $280 per kg of plutonium fuel produced.[39]

Results of the Pan Heuristics analysis differ significantly from those of other studies not because we use estimates of component costs far outside the range of costs in other studies, but rather because of our assessment that reprocessing costs are likely to be at the high end and uranium costs at the low end of the ranges of estimates used in other studies, and because we leave unspecified the costs of final storage of spent fuel and reprocessing wastes.

Uncertainties in the costs of uranium and plutonium fuels: The future price of uranium is uncertain, and the relative costs of uranium and plutonium fuels are quite sensitive to this price. The Pan Heuristics analysis assumes a price (in 1975 dollars) of $25 per pound of uranium oxide ($65 per kg of uranium metal), a price which seems high in the light of evidence available at present on uranium supplies and production costs. Our belief, and that of most experts in the field, is that the spot-market prices in Autumn 1976 of $40 per pound of uranium oxide reflected temporary shortages and will decline substantially as production expands. But, at sufficiently high uranium prices, plutonium fuel would become much more competitive with uranium fuel. Each rise of $10 per pound of uranium oxide reduces the relative cost of plutonium fuel by about $260; thus at $50 per pound of uranium oxide, plutonium and uranium fuel would cost about the same in our example, assuming that all other prices remain the same.

In addition, since about 5 kg of spent fuel must be reprocessed per kg of plutonium fuel, the cost of this fuel is quite sensitive to reprocessing charges. Although we believe reprocessing charges much below $300 (1976 dollars) cannot be realistically justified, some analyses use much lower estimates.[40] Lowering the assumed reprocessing charge by $100 per kg lowers the cost of plutonium fuel by $500 per kilogram.

It is possible that reprocessing costs will be much higher and uranium prices will be much lower than those used in the Pan Heuristics analysis. Uranium prices near or below $10 per pound and reprocessing charges in the vicinity of $500 per kg appear to be plausible possibilities. Of course, there are substantial uncertainties; thus investments in a plutonium

---

[39] In calculating overall savings in resources associated with implementing reprocessing, it is also necessary to include the effect on uranium prices. Because less uranium will need to be mined if reprocessing is implemented, uranium prices will be lower under this alternative. Based on the assumptions made in the various studies, inclusion of the effect of reprocessing on uranium prices would raise the net gain per kg of plutonium fuel by $200 in the ERDA analysis, $160 in the *GESMO* analysis, and $650 in the EEI study. Our own analysis of uranium supply indicates that the price effects would be much smaller than those assumed in these studies, especially in the EEI study. Further, most of these price savings would not occur until late in the period, when nuclear capacity is assumed to be large; thus they provide no justification for an early start on reprocessing.

[40] See, for example, U.S. Nuclear Regulatory Commission, *Final Generic Environmental Impact Statement on the Use of Recycled Plutonium in Mixed Oxide Fuels in Light Water Cooled Reactors (GESMO)* NUREG-0002 (Washington, D.C.: Office of Nuclear Safety and Safeguards, August 1976), Vol. IV, Chap. XI.

fuel cycle might prove to be profitable, but any prudent evaluation of such investments should also consider the possibility that they would prove to be far more unprofitable than the Pan Heuristics estimate presents here. If so, existing facilities might then be abandoned, as happened with the NFS plant at West Valley, New York. But, abandoning operation will not eliminate the liabilities for cleaning up radioactive wastes generated to that point in time.[41] Present uncertainties reinforce the case for deferring the decision to begin reprocessing until positive gains can be more confidently predicted.

Uncertainties in ultimate disposal costs for radioactive wastes: Regulatory concern with ultimate disposal of radioactive wastes is a relatively recent development in the nuclear field. Regulations are still in the formative stage, and no firm requirements have been set for many important aspects of waste management. Depending upon the eventual requirements, costs of ultimate disposal of radioactive wastes could vary by factors of 10 or more. Current governmental thinking appears to favour immediate isolation in geologic structures—probably salt beds or domes—but this is such a recent shift from the previous position in favour of "retrievable surface storage facilities" (RSSF), that no detailed conceptual designs have been costed.[42] Further, ERDA has been so uniformly opposed to direct disposal of spent fuel that no studies are available on conceptual designs for low-cost methods of accomplishing this.[43] In such a situation, meaningful estimates of the magnitude of the difference between the costs of disposing of spent fuel directly and the costs of disposing of radioactive wastes from reprocessing are not possible. For this reason, disposal costs in the two modes—reprocessing and no reprocessing—were left unspecified in our analysis. This is equivalent to the neutral assumption that disposal costs would be the same in the two modes. By contrast, other studies have used disposal cost figures which translate into savings for reprocessing of $240 to $360 per kg of plutonium fuel, about one half the net gain attributable to reprocessing in three of the studies.[44] The assumption of such large differences in final storage costs appears extremely questionable, not only because of the lack of an empirical basis, but also because a careful consideration of the heat output and volumes of wastes associated with reprocessing strongly suggests that reprocessing will raise the costs of final storage.

41 Estimates of the costs of dealing with the wastes at the NFS site range up to $540 million, compared to less than $20 million in total "lifetime" revenues: *Alternative Processes for Managing Commercial Existing High Level Wastes*, prepared for the U.S. Nuclear Regulatory Commission by Battelle Pacific Northwest Laboratory, NUREG-0043 (April 1976), Table 2.5, p. 12.

42 The EEI, ERDA, and *GESMO* studies all rely on estimates for the RSSF concept in developing costs of storage for high-level wastes from reprocessing.

43 The recent analysis of alternatives in waste management for the Nuclear Regulatory Commission states: "It has been assumed in the past that uranium and plutonium in spent fuel would be recovered and recycled. Therefore, detailed analyses of the technology of spent fuel disposal are not available in the literature": U.S. Nuclear Regulatory Commission, *Environmental Survey of the Reprocessing and Waste Management Portions of the LWR Fuel Cycle*, NUREG-0116 (Supp. 1 to WASH 1248) (October 1976), pp. 4–113.

44 See Taylor, V., *The Economics of Plutonium and Uranium* (Los Angeles: Pan Heuristics, 30 April, 1977).

Management of heat generated from wastes will be a major factor, perhaps the most important, in ultimate disposal charges. Previous federal plans for the RSSF would restrict the heat output of a standard waste container and assess storage charges on a per-container basis. This would make heat output the dominant influence on storage costs for high-level wastes.

Reprocessing spent uranium fuel would reduce its heat output at the time of transfer to final storage by about 7 to 15 per cent.,[45] but set against this reduction would be the two to four times greater heat output of the spent plutonium fuel produced in the reprocessing case. Thus, the relative heat output of materials transferred to final storage with and without reprocessing depends upon the ratio of plutonium to uranium fuel in the reprocessing case, which in turn depends upon the rate of growth in installed nuclear capacity. For the growth treated in *GESMO*, plutonium fuel averages 11 per cent. of the total fuel reprocessed. For this ratio, calculations based on *GESMO* data show that the heat output of reprocessing wastes sent to final storage would vary from about 106 to 114 per cent. of the heat output of unreprocessed uranium fuel sent directly to final storage. With lower nuclear growth, which seems quite probable, the relative heat output would be still higher.

With reprocessing, the original radioactive isotopes in spent fuel would contaminate much larger volumes of process chemicals, materials, and equipment in reprocessing and fuel fabrication facilities. Because of possible transuranic contamination, all these wastes would need to meet the same requirements of isolation and retrievability under probable federal regulations. The volumes which would need to be transferred to final storage are highly uncertain at this time because of uncertainties in regulatory requirements, the volume of materials which will be contaminated, and the reductions in volumes which economically might be achieved by compaction and incineration, the extent of secondary wastes generated by volume-reduction processes, and additional volume which might be added by waste space in packaging. Waste volumes with reprocessing might range from 24 to 130 cu ft per metric ton of fuel reprocessed compared to about 30 cu ft for one method of direct storage of spent fuel (Table III). At the lower end of this range, the volume of waste would be somewhat lower than the volume of encapsulated fuel rods. There are, however, substantial technical and economic uncertainties which apply to these operations in volume reduction. Most current estimates of reprocessing costs do not provide for extensive volume reduction operations. Also, if undertaken, such operations will themselves produce radioactive wastes. (These have not been included in Table III).

In summary, although lack of specification of regulatory requirements and the existence of numerous technical uncertainties preclude accurate

---

[45] Plutonium contributes this amount to heat output. The relative heat outputs of spent fuel and reprocessing wastes at the time they are transferred to final storage depends upon the time which has elapsed after discharge from the reactor. This time-lapse may differ for spent fuel and reprocessing wastes, depending upon the economics of temporary and final storage for the two waste forms. In what follows they are assumed to be transferred at the same time after discharge.

TABLE III

Volumes of Radioactive Waste Transferred to Final Storage
(Cu ft per Metric Tonne Heavy Metal)

Reprocessing Wastes	As Produced	After Packaging
Solidified Liquid Wastes	2–5 [a]	2–5
Fuel Cladding	12–20	4–20
Process Trash	50–100	13–38
Failed Process Equipment	2–14	1–23
Plutonium Fuel Fabrication Wastes		
All Types	21–83	4–43
Totals with Reprocessing	87–222	24–130
Direct Storage of Spent Fuel [b]		31

[a] Current regulations require solidification of liquid wastes; thus this is assumed for the output as produced.

[b] Direct Storage of Spent Fuel: internal Pan Heuristics analysis, based on encapsulation of 4 PWR fuel assemblies in a single zinc-filled container.

SOURCES: Reprocessing and Plutonium Fuel Fabrication Waste Volumes: Francis O'Hara, table compiled from industry survey data prepared for the Arms Control and Disarmament Agency, 8 August, 1976. Private communication from Richard Speier, Arms Control and Disarmament Agency, 23 August, 1976.

cost analyses of final storage costs at this time, both heat-output and waste-volume considerations suggest that such costs will be higher with reprocessing than with direct storage of spent fuel.

Possible savings from plutonium in perspective: Suppose, contrary to all reasonable expectations, that fuel cycle savings of 10 per cent. were possible per kg of plutonium fuel. What would this mean to the overall economics of nuclear-generated electricity? Nuclear reactors which start operating in the early 1980s will probably cost around $1,000 (1975 dollars) per kw of capacity. Operating at 60 per cent. of capacity and at a fixed charge rate of 15 per cent., the capital costs per kwh will equal 28 mills ($.028) per kwh. Operating and maintenance costs will total, perhaps, 3 mills per kwh; thus, non-fuel costs will equal about 31 mills per kwh. Fuel costs will, of course, depend upon uranium and enrichment prices. Consider prices which would produce a cost advantage of 10 per cent. for plutonium fuel, using other costs in the example of Figure 3. In Table IV, these costs are translated into per kwh fuel costs and added to the estimated non-fuel cost of 31 mills per kwh. The assumed fuel savings of 10 per cent. would lower total generating costs by 0·6 mills or 1·6 per cent., where plutonium was used. But, even if reprocessing were pursued to the maximum possible extent, plutonium fuel could provide less than 15 per cent. of total nuclear fuel requirements.[46] Further, generating costs

[46] For the growth rates currently being projected for nuclear power, recovered products will supply less than 25 per cent. of fissile-fuel needs. About one half of recovered fissile material is plutonium; the rest is recovered uranium. In calculating the cost of plutonium

are only about 50 per cent. of delivered electricity costs; thus the 10 per cent. cost advantage for plutonium fuel would reduce costs of the total system-delivered nuclear electricity by 50 per cent. of 15 per cent. of 1·6 per cent., or about one tenth of 1 per cent.

In using plutonium fuel, we would be risking the spread of nuclear weapons to dozens of additional countries for the sake of possible monetary savings which would be offset by a few months of capital-cost inflation

TABLE IV

Hypothetical Costs of Nuclear Electricity in the Early 1980s
(Prices in 1975 dollars)

	Uranium Fuel at $1,450 per kg (mills)	Plutonium Fuel at $1,320 per kg (mills)
Fuel Costs per Kwh		
Cost of fuel consumed [a]	6·1	5·6
Interest cost on fuel core [b]	1·1	1·0
Total fuel costs	7·2	6·6
Non-Fuel Costs per Kwh		
Capital costs	28	28
Operating and maintenance costs	3	3
	31	31
Total generating costs per kwh	38·2	37·6

[a] Assumes thermal efficiency of .33 and " burnup " of 30,000 MWD_{Th}–MTU, appropriate for a replacement load in a PWR.
[b] Assumes a core load of 80 kg of fuel per kw of capacity (typical for a PWR), a 10 per cent. carrying charge, 60 per cent. capacity factor, and an average in-core fuel value equal to one half the value of new fuel.

at recent rates. The word " possible " is stressed because it seems far more likely that early introduction of a plutonium fuel cycle would result in net economic losses rather than in net economic gains. Reprocessing will only be profitable at very high uranium prices, but it is likely that uranium prices will be quite moderate for at least the remainder of this century.

Uranium Scarcity: Myth or Reality? [47]

Intrinsic to arguments for the early introduction of plutonium fuels is the notion that uranium scarcity is upon us or looming just over the horizon. If we do not quickly introduce plutonium fuels, this argument goes,

fuel, credit was given for the recovered uranium; thus the (less than 15 per cent.) of fuel requirements met by plutonium fuel reflect the total monetary savings possible from recycle.
[47] This section presents, in extremely abbreviated form, the major findings of Taylor, V., *The Myth of Uranium Scarcity* (Los Angeles: Pan Heuristics, April 1977), referred to hereafter as, *Uranium Scarcity*. See that report for documentation and references.

known reserves of low-cost uranium will soon be exhausted, possibly bringing nuclear power to a complete halt or, at the least, driving nuclear fuel costs to high levels as we are forced to exploit marginal and very expensive sources of uranium. Analyses which " demonstrate " the economic advantages of introducing plutonium fuels use as evidence uranium prices which are far above historical levels to produce most or all of these advantages.

Arguments which depend on imminent exhaustion of resources of low-cost uranium are contradicted by official estimates, which show that quantities of such resources in known uranium districts are sufficient to meet expected requirements well beyond the end of the century. Many major new discoveries can be expected in the intervening years between now and the year 2000 if demand for uranium expands as anticipated. Further, the history of mineral extraction generally, as well as the brief history of uranium production in particular, indicates the likelihood that continuing improvements in technology will lower production costs, so what are now considered higher-cost resources seem likely to become low-cost resources in the future.

Forecasts that uranium prices will move steadily and rapidly upwards from " present prices " of $40 per pound of uranium oxide (U_3O_8) to $100 per pound or more reflect inadequate understanding of the economics of the uranium market. Exaggerated expectations for uranium prices can arise from a number of sources: first, by taking recent spot-market prices and trends as an indicator of expected prices in the long term; second, by taking estimates of " reserves " or " reasonably assured resources " as the best measures of the likely extent of uranium resources; third, by accepting official estimates of the growth of nuclear power as the basis for estimating the demand for uranium; and, finally, by ignoring the probability of continuing, cost-saving technological improvements in uranium exploration, mining and milling.

When these sources of exaggeration are understood and placed in proper perspective, continuing high uranium prices appear extremely unlikely. Rather, the need for uranium in this century appears capable of being met from resources which can be produced for less than $20 per pound, and prices under 10 dollars per pound in the first quarter of the twenty-first century seem at least a reasonable possibility.

Near-term uranium prices: A major source of confusion with respect to the outlook for uranium prices has been the rapid rise of prices in the spot-market. From a level of $6 per pound in 1973, prices of uranium oxide rose to $40 per pound in 1976, where they remained in early 1977. This movement has been interpreted by many to presage even higher prices in the future.

The first point to understand is that only a small fraction of uranium purchases occur on the spot-market, and thus these prices are a poor indicator of average uranium prices. On this deflated basis, long-term contract prices are below $15 per pound of uranium oxide through 1985 and, thus, are far below the commonly cited spot-price of $40 per pound. Of course, many of these contracts were negotiated when prices were much lower than today, but even new contracts for delivery in the 1980s appear to have been negotiated at prices (in 1975 dollars) below $20 per pound. Although contracts are being negotiated at prices well below the spot-market, the high spot-prices indicate very tight supply, with many buyers' needs still unsatisfied. High short-term demands, however, do not represent the true requirements of reactors for fuel, but rather reflect artificial demands created by past enrichment policies of the United States. These policies have forced the customers of the United States—who will buy most of the world uranium output over the next 10 years—to plan to deliver for enrichment quantities of uranium far in excess of their actual needs for enriched fuel.[48] As a result, these customers are rapidly building up excess inventories of nuclear fuel. The stage has been set for another cycle of " boom to bust " in the uranium industry, and the challenge now facing the United States Government is to modify its enrichment policies in order to minimise the harmful effects of the approaching downward trend of the cycle.

The simultaneous growth of uranium consumption and Reserves: Arguments that uranium is scarce and that we face future shortages and high prices often compare uranium Reserves and estimated future consumption to prove the case. What needs to be understood is that, by definition, Reserves refer only to uranium resources which have been specifically located and delineated by drilling and other engineering techniques. Placing resources into the " Reserve " category, thus, requires mining companies to make investments. These investments will not be profitable unless the Reserves can be mined in the relatively near future; therefore, current Reserves will look very small in relation to 25 years of growing production requirements.

The fallacy of measuring the adequacy of resources by Reserve figures is convincingly demonstrated by historical experience. In 1950, total United States Reserves of uranium oxide were estimated at 3,000 tons. In the next 10 years, 79,000 tons of uranium oxide were produced. In the single year of 1960, six times the 1950 Reserves were produced. But, rather than being exhausted by this production, Reserves were then estimated to total 187,000 tons, or 60 times the 1950 figure. In short, Reserves grew rather than diminished as consumption increased. This logical economic behaviour can be expected to continue in the future.

A supply curve based on ERDA resource estimates: The most commonly cited figures on United States uranium resources are the official estimates prepared by ERDA. Figure 4 shows a supply curve based on these estimates. ERDA resource estimates appear to understate the likely amounts of uranium available within the various forward cost categories. Further, the supply curve of Figure 4 assumes that mining and milling productivity remains fixed at 1975 levels, whereas historically, productivity has shown persistent improvement; thus this supply curve very likely overestimates the cost at

---

[48] For a complete analysis, see Taylor, V., *How the U.S. Government Created the Uranium Crisis (and the Coming Uranium Bust)*, (Los Angeles: Pan Heuristics, June 1977).

FIGURE 4

United States Uranium Supply Curve Based on Official ERDA Resource Estimates

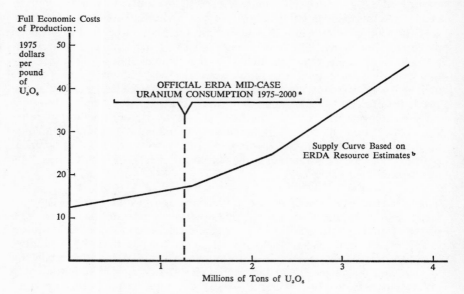

a Hanrahan, E., Williamson, R. and Bown, R., "World Requirements and Supply of Uranium," United States Energy Research and Development Association, presented at the Atomic Industrial Forum International Conference on Uranium, Geneva, Switzerland, 14 September, 1976; mid-case, no-reprocessing, .20 tails, implies about 1.3 million tons U308, Fig. 5, p. 19.

b United States Uranium Resources, 1 January, 1976 (reserves plus potential resources), from: Patterson, John A., Chief, Supply Evaluation Branch Division of Nuclear Fuel Supply and Production, " Uranium Supply Developments ", presented at the AIF Fuel Cycle Conference, Phoenix, Arizona, 22 March, 1976. Conversion of ERDA " forward costs " to full costs of production based on analyses by John Klemenic, Chief of the Supply Analysis Branch of ERDA's Grand Junction (Colorado) Office: See *Prepared Testimony of Dr. Vince Taylor for the California Energy Resources and Development Commission*, in the Matter of Generic Environmental Statement on Mixed Oxide Fuel (GESMO), U.S. Nuclear Regulatory Commission, Docket No. RM-50-5, 4 March, 1977, Addendum 2, pp. 2–1 to 2–6. Full costs of production are 1.5 to 1.7 times ERDA forward costs.

which uranium will be available in the future. Therefore, the lowness of the implied future uranium prices (in 1975 dollars) is especially note-worthy. Current official ERDA mid-case projections are that United States uranium consumption from 1975 until 2000 will be about 1.3 million tons of uranium oxide. The supply curve in Figure 4 shows uranium pro-duction costs of \$17 per pound of uranium oxide at this level of cumula-tive consumption. Our own exploratory projection of possible nuclear power growth—which is on the high side of what seems reasonable to expect on current evidence—implies cumulative consumption from 1975 to 2000 of about 0.8 million tons of uranium oxide, or a production cost in the year 2000 of about \$15 per pound.[49]

49 See Taylor, V., *Uranium Scarcity*, for a review of this evidence.

The longer-run outlook for the relative costs of uranium and plutonium:
Future productivity trends, which will be major determinants of the future
costs of uranium, are highly uncertain. From the inception of large-scale
uranium production until the early 1970s, unit costs (in constant dollars) of
uranium mining and milling in the United States appear to have declined
steadily at about 6 per cent. per year. If future costs (for ore of specified
characteristics) were to decline in the future at 2 per cent. per year,
uranium costs in the next century would be below $10 per pound.[50] Since
the early 1970s, however, uranium costs have risen significantly because of
(1) higher standards for environmental protection; (2) new requirements for
worker safety; (3) above-average wage inflation; and (4) a rise in the rate
of profit required by investors because of increased uncertainty about
nuclear prospects—the combination of which more than offsets continuing
improvements in production technology. These recent trends may well
continue, raising the possibility that future costs of production will be
higher than the estimates given above.

Although those who argue for reprocessing generally cite the likelihood
of continued upward pressure on the costs of uranium production as favourable to the economics of plutonium, the reverse is actually the case. Because
reprocessing, as compared to uranium production, requires protection
against levels of radioactivity many orders of magnitude higher, is much
more capital intensive, and requires construction to much higher standards
of reliability, the same fundamental factors which have raised the costs of
uranium production have had a far more devastating effect on the costs of
producing plutonium fuels. During the past five or six years, the costs of
producing uranium, after adjustment for general inflation, appear to have
approximately doubled, whereas estimated costs of reprocessing have risen
by a factor of more than 10.[51] Although further increases in standards for
environmental protection and worker safety may well cause another doub-
ling of the costs of uranium production, these same forces would raise the
costs of reprocessing far more dramatically, into thousands of dollars per
kilogram if recent history provides an accurate guide. With costs even of
$1,000 per kilogram, reprocessing would be unprofitable if uranium prices
were much below $200 per pound.

Recurring fears of uranium scarcity: The prevailing emphasis of United
States nuclear research on breeders and plutonium fuels is reminiscent of
the situation during and after the Second World War, when scientists
believed uranium to be so scarce that breeder reactors provided the only
possible means of generating significant amounts of electricity. This mistaken belief caused the United States to concentrate its efforts to develop
civilian nuclear power exclusively on plutonium breeders until the 1950s,
when the myth of uranium scarcity was buried under an avalanche of

50 *Ibid.*, section F.
51 In the early 1970s, reprocessing costs at the facility of Allied General Nuclear Services,
then under construction at Barnwell, North Carolina, were estimated to be about $300
per kilogram. Recent contracts offered by Cogema, the French reprocessing firm, had prices
(in 1977 dollars) of $500 per kilogram: *Nucleonics Week*, XVIII, 51, (15 December, 1977),
p. 15.

discoveries of uranium brought forth by the purchase programme of the
United States Atomic Energy Commission.

If the United States had not been simultaneously developing uranium-
fuelled reactors for naval use, the erroneous belief in uranium scarcity
would have delayed development of civilian nuclear power by perhaps a
decade. The early policy erred because it ignored the possibility that improvements in methods for finding and producing uranium would greatly
expand the supply of low-cost uranium. The irony is that the scientists
ignored the possibility of such improvements while assuming major
technical advances in the production and use of plutonium fuels.[52] We
should not repeat the same mistake again.

World uranium supply and demand: A possible concern is that, although
the outlook for supply and demand of uranium in the United States
appears reassuring, the same cannot be said for the rest of the world. A
survey of worldwide prospects for nuclear power growth and uranium
resources shows this concern to be unfounded. Indeed, the outlook, though
subject to greater uncertainty, seems likely to be more favourable than
that for the United States considered in isolation. This is because very
little of the remainder of the world has been seriously searched for
uranium. Yet, uranium districts already identified appear capable of meeting the expected needs of the non-communist world well beyond the year
2000.

Growth in nuclear capacity outside the United States: Official forecasts of
the growth of nuclear power outside the United States are far more
exaggerated than current official forecasts in the United States. For
example, the official nuclear targets for 1985 of the European Economic
Community—which includes most of the Western European countries,
except Spain, with plans for substantial nuclear capacity—were reduced
from 200 gigawatts (10^6 kw) in December 1974 to about 160 gigawatts in
early 1976, and are reported to be about to be reduced further to 125
gigawatts.[53] That even the reduced goals are optimistic is apparent from
an examination of actual progress in constructing nuclear facilities in the
countries of the European Economic Community. There is at present 19.4
gigawatts of nuclear capacity in service, a further 32 gigawatts under
construction, and in theory another 40 gigawatts planned for a total of
91 gigawatts, well below the target of 125 gigawatts. But this figure includes 17 gigawatts of planned additions for Italy, additions which are
considered extremely dubious in view of Italy's economic difficulties,

[52] Not only were the scientists wrong about the supply of uranium, but they also badly
underestimated progress in improving the utilisation of uranium in non-breeder reactors.
Between the 1950s and the early 1960s, the energy estimated to be obtainable per ton of
uranium consumed in converter (non-breeder) reactors rose by a factor of about six (to a
level approximately equal to that being achieved in current light water reactors). From a
resource standpoint, this is equivalent to a sixfold expansion of uranium supplies. See
Mullenbach, Phillip, *Civilian Nuclear Power* (New York : The Twentieth Century Fund,
1963), pp. 37–38. Future improvements in utilisation, such as those possible with heavy water
reactors, can be expected to reinforce likely technical advances in uranium production,
further reducing incentives for introducing plutonium fuels.

[53] *Nucleonics Week*, XVII, 41 (17 October, 1976), pp. 9–10.

although that country is refusing to allow the European Economic Community to lower its estimate.[54] Making this deduction, the upper limit for 1985 becomes 74 gigawatts of nuclear capacity.

It should be noted that the target estimates of the European Economic Community—and other countries—are used without change or qualification in estimates by OECD and IAEA of nuclear growth.[55] These estimates are often accepted internationally as the most authoritative estimate of international nuclear growth. Their limitations should be understood. A country-by-country analysis shows that official estimates, such as those by OECD and IAEA, uniformly overestimate the demand for electricity and underestimate the obstacles to rapid nuclear growth. As a basis for planning, such gross overestimates have led to many bad decisions. In particular, they have created an illusory uranium shortage which is being used to justify large investments in breeder technology, reprocessing, and recycling.

Although IAEA estimates are beginning to be reduced, their estimates of installed nuclear capacity for the year 2000 are still well above our own relatively optimistic projections. Their most recent estimate of installed capacity for the non-communist world—including the United States—in the year 2000 is 1500–1800 gigawatts (compared to only 69 gigawatts in 1975).[56] The political acceptability of such rapid installation of so many reactors seems very questionable, given the evident public opposition to the much more modest current installations. Further, electricity growth will fall well below the growth rates implicit in these forecasts. Our own rough estimates, which do not allow for the effects of increased political opposition to nuclear power, suggest that nuclear capacity of the non-communist world will total less than 1000 gigawatts in the year 2000.[57]

TABLE V

Projected Uranium Requirements to the Year 2000

Forecast Nuclear Capacity (GW)	Cumulative Uranium Requirements (thousands of tons of U_3O_8)	Annual Uranium Requirements
1,500–1,800 [a]	2,400–3,000	220–265
1,000 [b]	1,969	224

[a] Krymm, Rurik and Woite, Georg, " Estimate of Future Demand for Uranium and Nuclear Fuel Services ", *IAEA Bulletin*, XVIII, 5/6 (1976), pp. 6–15. Uranium requirements are based on recycle of uranium (not plutonium) after 1990, a tails assay of 0.25 per cent. and a load factor of 70 per cent.

[b] Uranium requirements assume no recycle of either uranium or plutonium, all light water reactors, 0.2 per cent. tails assay, and 70 per cent. load factor (author's calculations).

54 *Ibid.*

55 See *Uranium Resources, Production and Demand*, Joint Report of the OECD Nuclear Energy Agency and the International Atomic Energy Agency, 1975.

56 Krymm, Rurik and Woite, Georg, " Estimates of Future Demand for Uranium and Nuclear Fuel Cycle Services ", *IAEA Bulletin*, XVIII, 5/6 (1976), p. 7.

57 Europe: 300 gigawatts; United States: 340 gigawatts; Japan: 100 gigawatts; Rest of the world (non-communist): 100–200 gigawatts. See, Taylor, V., *Uranium Scarcity*, for further details.

Total nuclear power growth in the non-communist world and uranium consumption to the year 2000: The implications of the most recent nuclear forecasts of the IAEA and NEA for the adequacy of uranium resources are tabulated in Table V, which also shows the implications of more moderate nuclear growth.

There are great uncertainties about all aspects of future energy consumption and production. In the face of such uncertainties, there are substantial benefits to be gained by deferring commitments and investments until more is known about the probable course of events. The clear message is that the costs of deferring commitments to reprocessing and the plutonium breeder reactor are likely to be minor relative to the potential benefits. Unless growth proceeds much faster than seems likely on the evidence at hand, present investments in these nuclear activities will be costly and wasteful as well as dangerous. The prudent and rational course is to defer investments in these activities unless and until such time as the course of actual events provides reasonable assurance that they are needed.

V

THINKING THROUGH THE MIDDLE POWER NUCLEAR OPTION: THE CASE OF JAPAN

MUCH of our analysis so far has centred on the paths a country might follow in acquiring a few nuclear explosives, perhaps like the first implosion weapons, for a primitive force. Such a force might be used as a last resort against a similarly equipped or unequipped neighbour. It might employ little in the way of command, control and communications, and could exploit aircraft already at hand as the means of delivery; even commercial transport or passenger aircraft might do. A capability of this sort, unreliable and vulnerable to attack, might be as much of a liability to its possessor as a menace to other nations. Nonetheless it might be all that a less developed country aspired to, at least initially. In that case the main hurdle would be the difficulty in acquiring the highly concentrated fissile material needed for the bomb.

The picture of proliferation, however, if left at that, would be incomplete and misleading. An industrial power facing a more potent threat would be unlikely to find such a force adequate. It would need a much more sophisticated and varied arsenal of nuclear weapons, and the weapons would be only a small part of the sort of force it might aim to acquire. The difficulties a middle-sized industrial country might face in building a force which seemed adequate to its needs is worth investigating. Such a nation might find the acquisition of materials for weapons relatively easy, but not the construction of a dependable force which could be expected to withstand attack and retaliate effectively.

In contrast with the central role which the development of a secure capacity for an effective " second strike " has had in strategic analysis in the United States, treatments of smaller nuclear powers have given very little attention to this problem. When they have addressed it, they have made major methodological deficiencies, or they have used inappropriate assumptions about the performance of the attacking weapons, or else they have implicitly assumed a notable lack of determination or planning on the part of the attacker.

Japan is a good case to consider, because it undoubtedly has considerable technological skill. The Japanese space programme now costs over $200 million a year. This sum is four times larger than it was five years ago. In the Japanese fiscal year of 1977, seven satellites are scheduled to be launched by Japanese vehicles. Two launching vehicles are being developed—the " M " and " N " launchers. By 1977 the " M " launcher, a vehicle using solid fuel, is expected to be able to put about 60 kilograms into an orbit of 30,000 kilometres. The " N " launcher, the motors of which use liquid and solid fuels, will be able to launch 130 kilograms into geostationary orbit. Research and development are under way to improve rocket engines and to produce systems of guidance and control which can put several hundred kilograms into geostationary orbit. The technology of guidance is being developed by the National Aerospace Laboratory, Japan Aviation Electronics Industry, Nippon Electric Company, and possibly by other organisations. Japan Aviation Electronics Industry, which has been working on inertial guidance since 1965, is developing the guidance of the " N " rocket. Nippon Electric Company has also cooperated with the University of Tokyo on " real-time " radio-guidance. Efforts are under way now on communications, navigation, earth observation, geodetic, and other satellites. Japan is also developing ship nuclear propulsion. The technology in the ship *Mutsu* could lead to a future system of submarine propulsion.

Nuclear technology in Japan has so advanced that nuclear weapons could be constructed relatively simply. But the technology of destroying nuclear weapons has also advanced—techniques for attacking and destroying fixed targets, for locating concealed mobile targets, and for attacking bombers and missiles while on their way to target. Even if Japan were to attempt to maintain only a modest and secure capacity for retaliation against the Soviet Union with nuclear weapons, it could probably not do so with a modest investment in the production of such weapons, in view of the continually advancing means for destroying them which will be available to the Soviet Union. To some extent, the problems Japan would face are common to all middle-sized powers which might aspire to deter a much stronger nuclear power; to some extent, however, they are peculiar to the geographical situation of Japan.

Command, Control and Communications

By no means the least of the problems to be solved in designing a force capable of a retaliatory second strike is the development and integration of command posts, communication systems, and information-gathering systems and procedures into a network for reliably controlling the forces and launching them, if necessary, under suitable military and governmental control. This system must be able not only to evaluate and respond to attacks on the nuclear weapons and the means of their delivery, but also to survive attacks on itself. Such systems may be the weakest links in the British and French nuclear forces—a point which deserves further analysis.

The weapons systems which are considered the most impregnable—*i.e.*, missiles on submarines—present the most difficult problems in communications. In the case of Japan, these problems would be even more severe if the forces had to be deployed close to the European part of the Soviet Union—and, hence, far from the military and governmental centres charged with the responsibility and authority to order the forces to attack.

Distance to targets: For the purposes of attack by the Soviet Union on Japan, the two countries are close together. But for a Japanese attack on the Soviet Union vast distances separate them. A Japanese deterrent force capable of reaching only targets in the Asian parts of Russia would appear to have too little dissuasive power. Important as Vladivostok is as a port, it could be rebuilt if destroyed. The deterrent effect of a threat to its population—estimated at 426,000 in 1970—would be greatly weakened by the Soviet ability to threaten to destroy every major city on the Japanese mainland in return. (Tokyo has a population of almost 9 million and is in a metropolitan area of almost 22 million; Osaka has a population of 3 million in a metropolitan area totalling over 13 million; Nagoya has a population of 2 million in a metropolitan area of 3 million.)

By far the greatest concentration of Soviet population and industrial wealth is west of the Urals over 4,000 nautical miles away. Thus, the Japanese would have to reach to intercontinental distances with their nuclear deterrent forces.

Levels of damage: If the Japanese decided to build a force capable of retaliation in a second strike, how much and what type of destruction would they want this force to be able to inflict? The question cannot be answered definitely, of course; and yet attitudes on this matter would strongly influence the design of the force.

Professor Geoffrey Kemp conducted a survey of Soviet urban areas and estimated the warhead requirements for achieving various levels of damage in his study *Nuclear Forces for Medium Powers.*[1] He designated as damage level D_1 " the capacity to threaten in a second-strike mode the top ten Soviet cities *excluding* those protected by the Moscow ABM system permitted under SALT I (assumed to be Moscow and Gorkiy)." [2] For damage level D_2, the targets would be the 10 largest Soviet cities, including Moscow and Gorkiy.

We use Professor Kemp's estimate of the basic requirement for warheads capable of inflicting these levels of damage. The two levels do not differ very much in the number of warheads which must be delivered to targets. The main difference between D_1 and D_2 would be the number of warheads required to penetrate an antiballistic missile defence. Using warheads of one megaton and the requirement of " distributing 5 lbs per square inch over-pressure over the whole city ", Kemp estimates 26 delivered warheads would be required to achieve level of damage D_1 and 31 for level of damage D_2. We take a rounded value of 30 as being close enough in both cases. With 50 kiloton warheads, Professor Kemp estimates 181 would be required to achieve a level of damage D_1 and 220 for D_2; we use the rounded value of 200 for these cases.[3]

The total population of the cities on the target list for damage level D_1 is 15 million; for damage level D_2, it is 21 million.[4] Kemp estimates that about 50 per cent. of these would be immediate fatalities from the derived levels of attack.[5] Although his methods are admittedly very approximate, these results strike us as being reasonable enough for our purposes here, provided there has been no significant dispersal of the population before the attack.

[1] *Nuclear Forces for Medium Powers: Part I: Targets and Weapons Systems*, Adelphi Paper No. 106; and *Parts II and III: Strategic Requirements and Options*, Adelphi Paper No. 107 (London : The International Institute for Strategic Studies, Autumn 1974).

[2] *Ibid.*, Parts II and III, p. 4 (Italics in original).

[3] *Ibid.*, p. 26.

[4] *Ibid.*, Table I, Part II, p. 5.

[5] *Ibid.*, Part II, p. 5.

Some claim that much lower damage levels would be adequate. However, we do not believe that the Japanese, in planning a first-class nuclear force which they felt was adequate for their needs, would wish to settle for much less than the ability to inflict damage level D_1 or D_2.

Japanese Intercontinental Ballistic Missiles

Proponents of small forces of nuclear weapons argued in the early and mid-1960s that a moderate number of ballistic missiles in " hardened " sites would be virtually impossible to destroy. Their calculations generally assumed that the attack would be launched in a single salvo and that inaccuracy and unreliability would drive the required number of attacking missiles to several times the number of missiles attacked.

Today it is widely accepted that no " technological plateau " exists which guarantees that several missiles are needed to destroy one missile. But published calculations about " missile duels " still suggest that from 10 to 20 per cent. of the hardened missile force might nonetheless survive. However, as our calculations will show, a Russian nuclear attack against a Japanese force in the 1980s and beyond could be much more devastating than that.

Moreover, with advanced precision-guidance and munitions especially designed for hard targets, most of such a force might be destroyed without resorting to nuclear weapons. The present treatment, however, considers only attacks with nuclear weapons.

" Fratricide ": One incoming nuclear warhead can destroy another, or at least degrade its effectiveness. This effect is called " fratricide, " and many phenomena associated with nuclear detonations can cause it. If a second weapon is close to a target when a first weapon detonates in its vicinity, the second may be destroyed or rendered ineffective by the first's detonation. Or if the second arrives a little bit later relative to the first's detonation, the shock wave may deflect it from its target. Or rocks thrown up by the first detonation might destroy the incoming warhead. After a while, a high-altitude dust cloud will form from the first detonation, which may cause erosion of incoming high-speed reentry vehicles. " Time windows " between the various effects have led analysts in the United States to conclude that two ballistic missile reentry vehicles per target can be sent to " hardened " missile silos without one warhead either destroying or degrading the other. Because of the very narrow constraints of time, there may not be time to take account of knowledge of the success or failure of the launchers in the first wave or salvo of weapons to " reprogramme " weapons to targets in the second wave.[6]

Analysts have argued that the restraints necessary to avoid fratricide substantially increase the fraction of the United States intercontinental

---

[6] See U.S. Congress, Subcommittee on Arms Control, International Organizations and Security Agreements of Committee on Foreign Relations, Senate, *Analyses of Effects of Limited Nuclear Warfare*, 94th Congress, 1st Session (Washington : U.S. Government Printing Office, September 1975); and McGlinchey, Joseph J. and Seelig, Jakob W., " Why ICBMs Can Survive a Nuclear Attack ", *Air Force*, LVII, 9 (September 1974), pp. 82–85.

ballistic missile force which would survive a Soviet attack on their launch silos.[7]

Since fratricide appears to limit the weight of an attack per target which can be usefully applied, the phenomenon might be seen as an " equaliser " which guaranteed adequate survival of small, hard, launcher-based forces even from attacks by a power with much larger inventories of attacking forces.

However, results need not be so comforting for the smaller power. If we assume only two " time windows " per target, each of the two windows can have any number of weapons sent through it, but at most, only one will detonate effectively. Thus, the power of the attack can be increased beyond two reentry vehicles to compensate for unreliability, but it cannot compensate for inadequate accuracy.

Considerations of fratricide do have the effect, though, of placing a premium on attack systems with high probability of each reliable weapon destroying a target—*i.e.*, high probability of " terminal kill ". This in turn leads to a premium on high accuracy or high yield. Of these, high accuracy is the more important; halving the median miss distance has the same effect on the probability of destroying the target as an eightfold increase in yield.

Nominal characteristics of hypothetical land-based missiles: Two alternative nominal Japanese intercontinental ballistic missiles (ICBMs) are used in this analysis—the first with a single one-megaton warhead, the second a multiple warhead version with seven 50-kiloton warheads. Each version would have about the same destructiveness—about 30 payloads delivered to targets would achieve Kemp's damage level D_1 or D_2. In both cases, we assume the missiles have an overall reliability of 0·75 per cent.

Either would be in the " Minuteman class " in size, but exact sizes would depend on the success of the Japanese programme of warhead development.

The Japanese might go to the extra trouble of developing a missile with multiple warheads for two reasons. They might not wish to try developing a megaton-range warhead, particularly if they choose to test their weapons underground. Or they might desire the added capability to overwhelm an anti-ballistic missile system which the larger numbers of warheads of multiple warhead systems would give—either to penetrate Moscow's existing system, or as a hedge against an expanded system in the future.

For the blast resistance of the silos, we assume the same medium blast resistance attributed to Minuteman by the Defense Department in 1969—that is, 300 pounds per square inch (psi).[8]

The number of missiles which must survive a Soviet attack: To achieve damage level D_1—the 10 largest cities exclusive of those protected by the

---

[7] McGlinchey, J. J. and Seelig, J. W., *loc cit.*

[8] Plus all other effects associated with a detonation resulting in 300 psi at the target, including electromagnetic pulse and ground shock. The estimate of 300 psi is based on the chart " Threat to Minuteman ", presented by Deputy Secretary of Defense Packard to the Senate Armed Services Committee, 20 March, 1969. See U.S. Congress, Committee on Armed Services, *Authorization for Military Procurement*, 91st Congress, 1st Session, Part I (Washington, D.C.: U.S. Government Printing Office, 1969), p. 178.

existing Soviet system of antiballistic missiles—about 40 missiles would have to be able to survive an attack by the Soviet Union, so that about 30 reliable ones would reach their targets. To achieve damage level D_2, additional missiles would be needed to overwhelm Moscow's defences. If the Japanese were to plan on the basis of 100 warheads being intercepted by Soviet antiballistic missiles—the number of interceptors allowed in the current ABM treaty—an unreasonably large number of single-warhead intercontinental ballistic missiles would be required just to get through the defences. An additional 20 such missiles, approximately, would have to survive the Soviet attack in the multiple-warhead case.

Some illustrative calculations: The originally planned French missile force consisted of 27 medium-range ballistic missiles. Clearly the Japanese would have to deploy more than this number to have the capability to inflict damage levels D_1 or D_2 after the Soviet Union attacked. But calculations based on a force of this size to see how it might fare are illuminating.

Let us consider attacks on the Japanese intercontinental ballistic missiles in which the Russians used 5-megaton warheads delivered with an accuracy of $\frac{1}{4}$ nautical mile [9] with an overall reliability of 75 per cent. Such a performance is well within their capability today, and an even better performance could be achieved by the 1980s. If two weapons per target were used, then the survival probability per target would be about 7 per cent.[10] Thus, if the Japanese intercontinental ballistic missile launchers were equal in number to the original French medium range ballistic missile force— *i.e.*, 27 silos—then only about two of these would be expected to survive the attack and the expected number of reliable missiles which would be delivered to target would be about one and one half, provided the Japanese were able to reach the decision to launch before follow-on attacks arrived, and assuming they avoided the Moscow area with its antiballistic missile system.

The Soviet Union might allocate more warheads than the 54 implied above. If, for instance, four weapons per target were used—*i.e.*, with two per fratricide time window—the probability of survival per target would be only eight tenths of 1 per cent. Out of a force of 27 intercontinental ballistic missiles, the expected number surviving would be much less than one, practically zero—indeed it turns out to be two tenths of one missile.

In this situation, a more meaningful statistic is the 80 per cent. probability that no intercontinental ballistic missiles survive—that is, in this case the Soviet Union would have an 80 per cent. chance of killing all such missiles in the Japanese force. The chance that no more than one missile would survive is 98 per cent.

Of course, if the Soviet Union wished to increase the probability of success of the mission, they could allocate six weapons per target—162 in all— which would raise the probability of destroying all 27 missiles to 95 per cent.

Even if the Japanese had a force of 100 missiles, the Russians could still

---

[9] Median miss distance or circular error probable (CEP). With a 1,000-foot CEP, which the Soviet Union could achieve in the 1980s, the results quoted would hold against launchers hardened to 1,000 psi.

[10] Details of the calculations are to be found in appendix B below.

destroy most of it without overburdening their own resources. Then 200 warheads—which could come from much less than that number of boosters—would lead to an expected number of seven surviving Japanese intercontinental missiles, of which about five might prove reliable. Kemp's level D_1 requires six times as many. By using 300 warheads, the Soviet Union could drive the expected number of survivors down to two; and by using 400 warheads, the expected survivors could be reduced to less than one, 0·8 of a missile.

These 400 warheads would represent only a fraction of the total forces which the Soviet Union has deployed against the United States. They need not come out of SALT quotas, which deal only with intercontinental forces.

The Japanese would have to deploy a sizeable force to feel confident that they could inflict damage level D_1 or D_2. Against a force of only 500 attacking warheads, well within Soviet capabilities, they would need about 300 or more missiles in silos.[11] If the Japanese wished to be somewhat conservative, and made plans to meet a 1,500-warhead attacking force, they would need about 700 ballistic missiles—hardly a medium-sized force.

Our calculations are of only the simplest, crudest forms of attack, essentially salvo attacks with ballistic missiles in which all weapons arrive with such short warning and so close together in time that it would be physically impossible for the Japanese to launch their weapons before or during the attack. Using a combination of missiles and bombers, the Russians could substantially reduce the attacking forces required or greatly increase the probability of destroying every missile. It would seem foolhardy for the Japanese to assume that they could—and would—launch the intercontinental missiles which survived the first salvo in the short length of time before bombers could carry out armed reconnaissance attacks against the surviving silos. But if the Russians did not want to count on such delays, they could coordinate the armed reconnaissance missions with timed detonations designed to " pin down " the remaining missiles. The limited geographical area available to the Japanese for deploying their hypothetical ICBM force, and the close proximity of the Soviet territory to Japan, would argue that the Soviet Union could more easily mount such attacks against Japan than against the United States.

One final note on the political symbolism of the hypothetical Japanese intercontinental ballistic missile forces: the Japanese might hope to gain some political leverage by actually deploying such a force in hardened silos. But these effects could very well vanish or turn against the Japanese if, as this force was being gradually deployed, intelligence indicated that the Russians were constructing hardened, medium-range missile silos in Eastern Siberia, or even that they had been observed preparing launch sites for semi-mobile, intermediate-range missiles—an independent, proportional deterrent of Japan.

Japanese Submarine-Based Ballistic Missiles

Land-based ballistic missiles on hardened launchers do not seem to be

[11] *Ibid.*

a useful strategy for Japan. But what about the system with the greatest reputation for survivability—submarine-based ballistic missiles?

Range and transit requirements: Japan would probably not try to use, even initially, a missile with the range of the first American, British, and French systems. The range of these first missiles required them to operate in the Northeast Atlantic, or possibly the eastern end of the Mediterranean. For the Japanese, a transit of 13,000 nautical miles would be necessary to avoid the Panama Canal and go round South Africa.[12] Using a reasonable 12 knot average rate of forward movement, a round trip of 110 days would be needed to travel round South Africa and 90 days via the Panama Canal. To be " on-station "—that is, in position to fire on command—for just 25 per cent. of a cruise, therefore, would require cruises of 147 days for the South African route and 120 days for the Panama Canal route.

These should be compared to the nominal cruise time of 60 days for United States submarines. The Japanese would have to deal not only with such problems as crew endurance, but also with problems of maintenance at extended distances.

If they develop a missile with a range of 2,500 nautical miles—the published range of the Poseidon missile—the Japanese could use an operating area in the Arabian Sea approximately equal in area to that used in the northeast North Atlantic by the very first Polaris boats. Out of a 60-day operating cycle, roughly two thirds (41 days), however, would still be spent in transit. This could be reduced to 29 days if they used the Bay of Bengal—which would, however, require a missile with a range of approximately 3,400 nautical miles.

In all these cases, moreover, the submarines would have to traverse the waters in and around Indonesia—in our calculations, we have assumed they use the shortest route, through the Straits of Malacca. The Straits of Malacca are territorial waters, although the right of innocent passage is recognised in international law. This route requires travelling about 1,000 nautical miles each way in water less than 100 fathoms deep—the minimum depth used by American ballistic missile submarines (SSBNs) for submerging—including about 500 nautical miles in the narrow Straits.[13]

It would be ironic if Japan developed a nuclear deterrent force dependent upon these sea lanes, since a major concern of Japan's defence is her dependence on these sea lanes for crude oil: 87 per cent. of Japanese crude oil comes from the Middle East.[14]

It is common practice for hostile antisubmarine warfare (ASW) forces to wait off submarine ports and attempt to pick up the trail of the submarines as they depart. To counter such " port-watch-and-trail " tactics, various " delousing " techniques are used. For high confidence, delousing usually involves the assistance of friendly forces—aircraft, surface ships,

[12] We assume that the alternative of forward basing of submarines in the Atlantic and movement of crews by air is not open to Japan.

[13] See Miller, Richard A., " Indonesia's Archipelago Doctrine and Japan's Jugular ", *U.S. Naval Institute Proceedings* (October 1972), pp. 27–33, for discussion of the various passages through Indonesia and the international status of Indonesia's claims on inland waters.

[14] Endicott, John E., *Japan's Nuclear Option: Political, Technical, and Strategic Factors* (New York: Praeger Publishers, 1975), p. 31.

or submarines. These would help both break the trail and verify that it has been broken. Such assistance may be essential if the antisubmarine warfare forces in question have the technological or numerical edge over the evading submarine. If the Japanese were to attempt to use the Arabian Sea or the Bay of Bengal as operating areas, they might also have to deploy forces for delousing in the Indian Ocean, because of the possibility of Soviet antisubmarine forces intercepting the SSBNs as they left the Straits of Malacca or whatever passage they used through the Indonesian archipelago.

In addition to the operational problems associated with such long passages through uncertain waters, the Japanese would face a severe problem in communicating with submarines so far away.

Japan appears to be deceptively close to European Russia via the Arctic Ocean. However, despite enthusiasm about using the " Northwest Passage " over North America by early proponents for the nuclear submarine, Arctic passages remain impractical for routine, all-season operations. The first difficulty the Japanese submarine attempting such passage would face would be crossing the Bering Sea and Bering Straits. These are shallow waters, making manoeuvres between the ice above and the seabed below hazardous. The *Nautilus* was forced to turn back on its first attempt to penetrate the Bering Sea; not until a later trip in August was it able to find its way.

In case of an emergency on board, one of the first actions a submarine commander is likely to want to take is to surface. This can prove difficult in the Arctic, where he may have to search for many hours to look for a polyna—a clear break in the ice—able to accommodate his vessel, and then hope that the shifting ice will not abruptly close the polyna, forcing a hasty submergence.

A less dramatic problem of polar operations relates to the inertial navigation systems of submarines. These depend on gyroscopes, which in turn depend on the earth's rotation, and they become poorer and poorer as the North Pole is approached.

Thus, if Japan were to attempt to build a sea-based missile deterrent force, it would initially require missiles with a range of at least 2,500 nautical miles. Unless a force using missiles with a range of 4,000 nautical miles or more were developed, operations would be hampered by problems associated with the use of operating areas far removed from the home base and, for practical purposes, accessible only through the Indonesian straits. Only now, with the Trident I and the SS-N-8, are the United States and Russia putting missiles of this range on their submarines.

The technology of antisubmarine warfare: Submarine-based ballistic missile systems enjoy a reputation of near invulnerability to attack from antisubmarine warfare forces. Professor Kemp, for instance, states, " The overwhelming consensus of opinions in the published literature is that, while there have indeed been technological breakthroughs in ASW techniques, the advantages still favour the hunted rather than the hunter." [15]

---

[15] Kemp, G., *op. cit.*, Part I, p. 20.

But this assessment assumes combat between two equally advanced technologies.

The success of submarine detection and trailing techniques depends on relative technical advantage. For instance, in the case of one submarine hunting and attempting to trail another, using passive sonar, success is critically dependent on which submarine has the greater sonar range. If the seeker has the greater reach, then he has a high probability of being able to detect the hunted submarine without being detected himself, and a good chance of successfully manoeuvring behind the hunted submarine. Once behind, it is easier for him to keep contact and harder for the submarine under trail to detect his stalker. With sufficient range advantage, the seeker then can often maintain the trail for several days without detection.[16] Even if the hider knows or suspects he is being trailed, evasion tactics are much less likely to succeed if the range advantage lies with the seeker.

Range advantage depends not only on the relative qualities of the sonars of the two boats, but also on their relative noise levels. Thus, to deny significant range advantage to Russian attack submarines, the Japanese would have to approach the Russian level of 1980s technology in both passive sonars and quieting of submarines.

The Japanese would have to do better than achieve the technical capabilities possessed by the United States or the Soviet Union in the 1970s. Japanese boats in 1985 would have to contend with Russian technology in 1985. And as the Russian boats—and other systems of antisubmarine warfare—continued to improve to, say, the technological advances of 1990, the Japanese would have to continue to modernise their force.

Variability and choice: If the Japanese set out to design a true " second-strike " force, it would not be enough for the force to survive the test of unimaginative attacks by half-hearted Russians attacking under expected or average conditions. Instead, they would have to design against resourceful and determined attackers, who showed military good sense and who chose to attack under conditions favourable to the attack. Designing and operating a force so that no weak spots are presented can make the task much more difficult and expensive, but that is what is required. For a force based on a small number of submarines, deviations from average situations can be quite large.

Consider, for example, the number of submarines on station. Here the British experience is instructive. Britain has four " Resolution " class SSBNs. According to Ian Smart [17]: " Major refits for each submarine should occur after every three years of operation, with the target time for a major refit being six months." [18] Even if a major refit took 12 months, the boats would have an in-commission rate of 75 per cent. In addition, the British

---

[16] How is trail ever lost under such circumstances? Equipment malfunction, of course, is one possibility. But intrinsic to the sea as a medium is the great variability of conditions in acoustic propagation. If conditions worsen, a trail may be lost. If they improve, the covert trailer may be detected. Changes may occur at one ship's sonar frequencies but not at the others, or a local source of noise may cause the trailer to lose contact.

[17] See *Future Conditional: The Prospect for Anglo-French Nuclear Cooperation*, Smart, Ian, Adelphi Paper No. 78 (August 1971), p. 15.

[18] *Ibid.*

plan is to have their boats carry out four patrols a year, each lasting 8 to 10 weeks. Using an average of nine weeks yields a fraction of time on patrol of slightly over two thirds for boats in commission. The average number of boats on patrol would then be at least $2(4 \times 0.75 \times 2/3)$. Yet Britain " has had to concede that it will have only one on station at some times." [19]

Implicit in this British concern over deviations from the average number of boats on patrol is the realisation that a potential attacker will, if possible, choose to attack when his opponent is weakest, and not under " expected " conditions.

Another highly variable factor is the acoustic conditions for sonar search. These vary greatly with the state of the sea and the water temperature gradient, among many factors. It might thus be difficult to keep track of a submarine all of the time, or under typical conditions, but nonetheless conditions for location will be very good during occasional periods. Consider the North Indian Ocean which is of interest because it is the probable operating area for a Japanese submarine-based nuclear missile force, if the missiles have a range less than 4,000 nautical miles. These seas are usually calm during the winter, the time of the northeast monsoon season. For instance, wave heights are then less than 4 feet about 80 per cent. of the time, as opposed to 30 per cent. of the time in the summer. Since rough seas generate noise which can degrade performance of sonars operating at higher frequencies,[20] the Soviet Union could expect a greater-than-average likelihood of success if it conducted an " exercise " against Japanese submarines in these waters in the winter.

Finally, the Russians might choose to exploit a " search-and-hold " strategy. They would not simply set out to destroy as many Japanese submarines as they could, as fast as they could. Instead, they would begin searching for submarines and trailing them when found. The actual number of submarines under trail would fluctuate, of course—depending on changes in environmental conditions, operating practices of the Japanese fleet, and luck. The Soviets would delay the actual attack of the submarines until the combination of these factors caused the number of submarines under trail to exceed a certain threshold, and then attack.

If a large number of submarines are on station, as is the case for the United States fleet, then the hunting nation might have to wait an unreasonably long time before all, or all but one or two of them, are simultaneously under trail. However, the situation is much different in the case of the three, two, or even one submarine which a medium-sized country might have on station.

Just as the deployment of a " counterforce " capability against the Japanese land-based intercontinental ballistic missile force might undermine whatever political benefits the Japanese hope to gain from that force, so

[19] *Ibid*.

[20] " It is clear that the rough sea surface, not too far from the location of the measurement hydrophone, is the dominant noise source at frequencies between 1 and 30 kHz (kilohertz) ": Urick, Robert J., *Principles of Underwater Sound for Engineers* (New York: McGraw-Hill, 1967), p. 166. At low frequencies—say, below about 300 hertz—surface shipping is the dominant source of noise.

also would Russian antisubmarine warfare exercises which successfully
found all or most of a Japanese submarine fleet at sea.

Some examples: Our calculations are greatly simplified for two reasons.
First, modelling antisubmarine warfare operations is complicated, and we
wish to keep these calculations clear. Second, public information on the
capabilities of antisubmarine warfare which are sufficiently quantitative
for use in calculations is sparse.

Consider the case of two ballistic missile submarines on patrol, which
the Russians attempt to find with attack submarines. The Russians could
conceivably search for the ballistic missile submarines and simply attack
the first as soon as they find it, hoping that its destruction will go undetected,
or that the Japanese are uncertain of the cause of its loss—this case assumes,
of course, an attack with non-nuclear weapons—at least for the length of time
required to find the other submarine, or that the Japanese would be deterred
from hasty use of the one remaining. We assume, however, that the Russians
do not attack until both submarines have been found, and that they then
attack with sufficient force for each submarine to have virtually no chance of
surviving to launch its missiles. It is thus necessary to hold trail on the
first submarine found while the search continues for the second.

We need estimates of two parameters related to the effectiveness of searching and trailing, respectively. The first is the effective sweep rate for each
attack submarine in the search. This can be thought of as the average
ocean area searched per unit of time. For our calculations, which assume
the attack submarines have substantial technological advantage in sonars
and quieting, we use an effective sweep rate of 90 square nautical miles
per hour per attack submarine. This can be visualised as an attack submarine travelling at 15 knots with a detection range of 3 nautical miles.
Or the searching submarine could be travelling at $7\frac{1}{2}$ knots, with a
capability of detection which extends to 6 nautical miles.[21]

The capability of an antisubmarine warfare unit at trailing is measured
by the expected length of time the unit can hold trail—its " mean holding
time ". We assume a mean holding time of eight days, an assumption consistent with the view that the Russian units would have substantial technical advantages over the Japanese units in sonars and submarine quieting
techniques.

If we now specify the size of the operating area of the ballistic missile
submarines, the number of attack submarines engaged in the search, and
the number of missile submarines under trail at the beginning of the search,
we can calculate the probability that both missile submarines will have
simultaneously been under trail at least once as a function of how long
the search and trail effort has gone on [22] (See Table I).

The case of an operating area of 200,000 square nautical miles is of
interest both because an area about that large was available to the first

---

21 In fact, the probability of detection is a function which varies continuously with
distance. The sweep rate is the average speed of the submarine times the integral over
distance of the distance times probability of detection.

22 Our calculations use a simple Markov process solved by computer. Mathematical
details appear in appendix B below.

American ballistic missile submarines, and because an operating area of about that size could be achieved in the Arabian Sea with a Poseidon-range (2,500 nautical miles) missile with the requirement that the submarines be within range of Moscow.

Although the early Polaris submarines were quite secure, a country of medium size which attempted to operate in such a constricted area today without substantial technical advantages would be vulnerable to attack. A search operation using 30 attack submarines would have an 80 per cent. probability of having both missile submarines simultaneously under trail at least once within eight days of initiating the search, assuming that initially none were under trail. They would have a 97 per cent. probability of success by the end of 16 days.

The Soviet Union would pay no major penalty for failure or for taking an extended period to find the ballistic missile submarines. (Although the time available will depend on the nature of the confrontation between the two countries, it can also be said that the nature of the confrontation may well be determined in part by estimates of the time required to find the submarines.)

TABLE I

Probability of Simultaneously Trailing Two Ballistic Missile Submarines (SSBNs) at Least Once in a Given Elapsed Time

Effective sweep-rate per attack submarine: 90 nmi^2/hr
Mean holding time: 8 days
Initial conditions: No SSBNs Under Trail

Operating Area (nmi^2)	Number of Search SSNs	Elapsed Time (days)	Probability Both Simultaneously Located at Least Once
200,000	15	8	0·44
		16	0·74
		32	0·95
200,000	30	8	0·80
		16	0·97
		32	>0·99
400,000	15	16	0·38
		32	0·66
400,000	30	8	0·45
		16	0·75
		32	0·95
800,000	30	16	0·31
		32	0·57
800,000	50	16	0·66
		32	0·91

One can also calculate the average number of missile submarines under trail at any one time assuming the attack submarines merely search and hold trail, but do not attack, and assuming the process has had a chance to settle down from the initial conditions. For instance, in the case of 50 attack submarines [23] searching 800,000 nautical square miles, the average under trail is about one—*i.e.*, the operation of searching and trailing has an average effectiveness of one half. However, this sort of " average " effectiveness can be misleading. The 60 per cent. chance that both boats can be simultaneously found within 16 days, and the 91 per cent. chance they can be found within 32 days has more meaning.

The case of only one submarine on patrol is also important for ballistic missile submarine forces of medium size. As noted above, even with a force of four vessels, each with two full crews and a favourable geographical location, the British apparently must operate at least some of the time with only one submarine on patrol. Accidents or unscheduled maintenance or delays in port could lead to situations where even a larger force would have only one submarine on patrol at some time. Thus, the case for one missile submarine deserves attention.

For one patrol boat, the probability of success in the search by time " t " is given by a simple exponential:

$$P(t) = 1 - e^{-mst/A}$$

in which " m " is the number of search units, " s " is the sweep rate per search unit, " A " is the area to be searched, and " t " is the time since the beginning of the search.

The expected length of time before the search is successful can also be expressed simply; it turns out to be the reciprocal of the coefficient of " t ":

$$\text{Mean search time} = \frac{A}{ms}$$

Consider again the case of a 200,000 nm² operating area. Using 15 attack submarines for the search,

$$\text{Mean search time} = \frac{200,000}{(15)\cdot(90)} \text{ hours,}$$
$$= 148 \text{ hours,}$$
$$= 6\cdot2 \text{ days.}$$

The probability of success would be 92 per cent. after 16 days.

We have assumed no *a priori* information on the location of the hunted vessels, at the outset of the search, other than that they are in the operating area. In practice, the Japanese would probably find it hard to achieve this. For instance, the Soviet Union could begin search when a submarine has

---

[23] Capt. John E. Moore, editor of *Jane's Fighting Ships*, writing in the Foreword to the 1974–75 edition, estimates that by the early 1980s the Soviet Union would have some 120 to 150 attack submarines at current building rates. See *Jane's Fighting Ships 1974–1975* (New York: Franklin Watts, Inc., 1974), p. 75.

just gone on patrol, so its initial location would be better known than is assumed above.

Out-of-range patrol areas: If the ocean area from which targets were in range were too constricted, the Japanese might put their submarines on patrol out of range of targets. They could then be ordered to move forward in a crisis. They would have to inch forward slowly, however, certainly at no greater speed than 10 knots or so, if they were to try to avoid detection. The slower they moved, the more time the Russians would have to hunt and attack them, evacuate cities, and place pressure on the Japanese government and populace to draw back. Such extended games of nerve do not favour the substantially smaller power.

Operations in the North Pacific: With a missile having a range of 4,000 nautical miles or more—such as the Trident I, the fifth generation of American submarine launched ballistic missiles—the Japanese would have large operating areas available close to home. Unfortunately, they would also be close to the Soviet Union. Numerous systems for open-ocean searches over a wide area or for extended trailing should be examined for this case. Possibilities are a Soviet SOSUS system for reducing the search area or setting up picket lines; the use of antisubmarine warfare by air attack; and trailing techniques using multiple types such as SSNs assisted by aircraft and surface vessels. A major determinant, again, would be how successfully the Japanese reduced noise and any other observable factors which Soviet sensors of the 1980s and beyond might be able to detect.

Other Systems

Other potential systems include long-range cruise missiles, probably based either on submarines or surface ships. For a given range and payload, these would be smaller than ballistic missiles and so would have some advantages in deployment over ballistic missiles. The biggest question for the Japanese would be whether they could assure themselves that such a system would give them adequate air defence penetration far into the future. (There is some controversy in this country over whether stand-off cruise missiles would be effective against Soviet defences of the 1980s.)

Perhaps we should emphasise that we are not contending that the Japanese could not develop and maintain a modest second-strike retaliatory capability. What we are suggesting is that such a force would not be both cheap and dependable—that reaching Trinity may be only the beginning of a long, expensive and irreversible journey.

LIFE IN A NUCLEAR ARMED CROWD

MUCH of the literature on the spread of nuclear weapons views a world with many nuclear powers as vastly more dangerous than today's world. Commentators outside the United States have not always viewed the prospect as gloomily as have Americans. To General Pierre Gallois, an early expositor of the view that one country could not be counted on to commit suicide to protect another, it is natural for each state to possess these most powerful weapons; he believes it would have a stabilising effect. Countries of the " third world " think that the wider distribution of nuclear weapons is a requirement of distributional justice. Military nuclear power, along with the world flow of income and stock of capital, should, they think, be shared more equally. Not all the advocates of nuclear weapons in the " third world " hold that every country should have nuclear weapons; many concentrate on the needs of their own country. Security needs—or perhaps opportunities—are perceived with respect to powerful adversaries armed only with non-nuclear weapons; more urgent concerns are generated by adversaries who have or might get nuclear weapons. Hedges against the breakdown of alliances are sought. And nuclear weapons sometimes have symbolic worth. British prime ministers have seen them as " entry tickets " to great power negotiations; so have Argentinian parliamentarians.

Despite the existence of much rhetoric justifying their acquisition, few of the countries with the capacity to make these weapons have done so. They have not acquired nuclear weapons because of high costs, internal political opposition, concern about reactions from neighbouring countries, an absence of perceived military threats and a belief in the adequacy of the guarantee provided by alliances. As a result, the entry rate into the category of countries possessing nuclear weapons has been low, far lower than has often been predicted.

Two developments now promise fundamental changes: one is the growth in civilian nuclear programmes and therefore an increased capacity to acquire nuclear weapons cheaply and rapidly; the other is a weakening of confidence in American guarantees of protection of allies.

By 1985, about 40 countries will have enough fissile material to make three bombs or more; almost as many are likely to have enough fissile material for from 30 to 60 weapons or more. From 1985 onwards, many more countries will be in a position to acquire nuclear weapons cheaply and rapidly.

Concern about reliance on protection by the United States has been widely expressed abroad in the past several years by officials and private persons in many countries. So far this factor apparently has not led governments in countries which are most firmly parts of the system of alliances with the United States, namely the countries of Western Europe and Japan, to acquire weapons. In contrast, most of the countries which now appear to be moving towards a capacity to produce weapons are either outside any system of alliance, are on the fringes of the American system of alliances, or are drifting out of the American system. They read ominous signs in the collapse of the American position in Southeast Asia, in Congressional opposition to support for South Vietnam and Turkey, attempts to block the development of an air base in the Indian Ocean, and support for withdrawals in the Republic of Korea. Moreover, the growing military power of the Soviet Union, together with an evident American preoccupation with reducing the likelihood of a nuclear exchange between the United States and the Soviet Union, are not altogether reassuring to countries which have been dependent on the United States. One response is to take out " nuclear insurance ".

The existence of this " overhang " of countries capable of making weapons means that the great powers will have to allow for the possibility that smaller ones will respond to certain of their actions by acquiring nuclear weapons. Events outside their control could have this effect anyway. Although this concern, especially about Japan and Germany, has not been absent in the past in American thought on foreign policy, it is now moving to the forefront.

In contemplating the world in the late 1980s and beyond, one of the most important questions to ask is: How likely is it that the countries able to acquire nuclear weapons rapidly and at a small increment in cost will do so? Analogies with " chain reactions ", " branching processes ", and " super-saturated solutions " suggest an exponential rate of growth in the future in contrast with the low, linear rate of the spread of nuclear weapons since 1945. In retrospect, it may appear that the Indian nuclear explosion of 1974 was a " triggering " event between a linear past and an exponential future.

Although there is a real chance that many countries will take the additional step and acquire weapons, it is not certain. There exist contradictory forces which may substantially moderate the rate of acquisition of nuclear weapons for the next decade and longer.

Let us however suppose that the " chain reaction " proceeds and that many additional countries acquire at least a few nuclear weapons by 1990. If this is so, we should consider several questions. They are: What is likely to be the stability of a system of countries with a few large, some middle-sized and many small powers with nuclear weapons, or which are close to having them, together with many more countries which do not have them? What shocks to the international system might occur and what equilibrating mechanisms might exist or be developed? What effect would these developments have on the likelihood of war, non-nuclear or nuclear? What consequences of the use of nuclear weapons are to be expected? How will relations within alliances affect decisions on acquisition, and, conversely, how is the wider possession of nuclear weapons likely to affect alliances? What new nuclear threats does this pose to American armed forces or territory? What might the United States have to do to survive in a world with many possessors of nuclear weapons?

The Stability of a World of Proliferation

Worries about stability in a world in which nuclear weapons are spreading rapidly stem from the facts that a great many rivalries and hostilities exist among—and within—countries; that the acquisition of nuclear weapons may drastically alter some relations among the powers and have wider repercussions short of war; and that it might increase the likelihood of wars and, more importantly, their destructiveness.

Short of war, changes in status or relationships among the powers might lead to coercive threats by the new possessors of nuclear weapons; internal political adaptations in response to new external threats; withdrawal of guarantees previously given by great powers in connection with alliances; creation of sources of nuclear weapons for states and sub-national groups; increases in defensive measures such as civil defence, active defence, and defence against terrorists; and threats of use by non-governmental groups, possibly supported by outside powers.

Some of these changes might lead to nuclear war between the two great powers; nuclear war between a great power and a smaller one; nuclear war between two small powers; or a general nuclear conflagration involving many powers of different strength. These changes might also lead to wars using conventional weapons involving the great powers and taking place under the " umbrella " of nuclear threats; or non-nuclear war limited to small powers, some of which may possess nuclear weapons and some of which may not.

In general, not all the probabilities of undesired consequences can be reduced simultaneously because some are conflicting. For example, if the great powers offer protection to allies so as to reduce the likelihood that the latter will be coerced or attacked, then the great powers will have to accept some risk of conflict among themselves. (Perhaps the choice would be between some risk now and a greater risk later.) A country might decide to be unaligned and have little armament in order to reduce the risk of involvement in conflict, but it would then have to pay the price of making unwelcome changes in its external or internal policies.

The " Two-Country Model "

With these two possible consequences in mind we can examine a number of simple cases. These include the standard " two-country model " relationship of the United States and the Soviet Union in which it is usually assumed, implicitly or explicitly, that the two countries have no allies or, more generally, no highly important conflicting interests in third areas. Reducing the probability of nuclear war as close to zero as possible is taken to be the overriding objective of each government. Another assumption of this " model " is that only the prospect of civil devastation is stabilising. An ability to defend oneself against nuclear attack or to exhaust the adversary's nuclear forces is worse than useless, because it stimulates " unlimited " increases in the level of nuclear forces and expenditures by the two countries and its creates incentives for pre-emptive attack. Hence, according to this " model " at least one and preferably each of the governments should commit itself to a policy of national suicide. Limited nuclear attacks or responses should not be planned. Finally, it assumes that any use of nuclear weapons will develop—with high probability—into a complete nuclear exchange.

The standard function of this model has been to provide analytic support in favour of common reductions in their nuclear forces by the United States and the Soviet Union—or just the United States unilaterally.

Even in the abstract binary world which it assumes, this model has crippling defects. For example, it ignores the constraints of limited resources, which are sufficient by themselves to limit the competition in armaments. Another defect is the assumption that escalation is inevitable and that massive response to limited attacks requires that governments be committed to homicide and suicide. This is not impossible but it is hardly a plausible assumption for most governments. Alternatively, the postulated responses are not to be taken seriously; deterrence alone is the objective. In the model which we are discussing the assumption is made that there are no external interests worth any risk of nuclear war, yet any damage from a nuclear atack within one's own territory warrants a wholly suicidal response. This attributes to the state a narrowness of interests which would be rejected by most supporters of democracy if these assumptions were badly expressed outside the context of nuclear deterrence. Finally, the model is inconsistent with the behaviour revealed by governments repeatedly since the advent of nuclear weapons. Most policies of the United States and the Soviet Union, including those on nuclear weapons, have been directed towards the support of allies and the deterrence of adversaries in third areas. That is where the great powers come into conflict and where the gains and losses in the competition between them occur.

Therefore, even when analysing the behaviour of the two great powers in this schematic way, it is essential to include their interests in third areas. When this is done, more complex and realistic types of behaviour become apparent. The limited uses of force can occur between small countries with the support of the great powers, including their involvement in combat. There might even be conflict directly between the great powers with conventional weapons or with nuclear weapons which is focused on and limited to a third area. The possibility of the selective and limited use of conventional or even nuclear weapons against targets in their respective territories cannot be excluded as an outgrowth of a conflict in a third area.

In sum, in a world in which there are important interests in third countries, *i.e.*, in the real world, there cannot be exclusive concentration by the two great powers on deterring attack on each other's territories without eliminating the guarantees which they have given to third countries.

Moreover, there is an important spur to proliferation implicit in this model: the absolute avoidance of conflict between the two powers requires the withdrawal of guarantees, including nuclear guarantees, from countries where serious great power rivalry exists. This, in turn, risks increasing perceived vulnerabilities and the incentives of third countries to acquire nuclear weapons.

Large Country versus Small Country

What about the case in which the adversaries differ greatly in size? A number of countries which could have nuclear weapons in 1990 would have only a small fraction of the forces of the United States or the Soviet

Union. Would this make the small power effectively equal to the large one? On the view that nuclear weapons would—and should—be used against civilian populations, a small power might appear about as strong as a large one at only, let us say, 1 per cent. of the cost. However, the larger country would have many more resources for gathering information, for command and control, for civil defence, and for anti-missile and air defences. With the precision of delivery which at least the great powers will possess by the late 1980s, fixed bases, including hardened facilities, will be extremely vulnerable. Mobile missiles, based on land which can be roughly located, can be attacked with missiles able to cover large areas with high blast over-pressure. Even the submarine forces of powers of moderate size may have serious problems of survival.

As a result, large countries, in general, will have the capacity to destroy smaller ones but not vice versa. This difference in destructiveness implies a difference in bargaining power. To be sure, the smaller country might be able to damage the larger one; it may be able, as President de Gaulle put it, to " tear off an arm ". In this formulation, the stakes at issue for the larger country are not supposed to be comparable to the damage, severe but less than total, which the smaller country might inflict. But what about the expected damage to the smaller one in relation to its stakes? Few alternatives are so grim that the prospect of nuclear annihilation would be preferred. Perhaps Israel, faced with a " final solution " by Arab states comes closer to providing such an example than any other in prospect. (But Israel faces, most immediately, not a great power but several small ones, and it has the support of a great power.)

At this stage, the principal inferences which can be made about this case are that the possession of nuclear weapons by the smaller power is likely to induce caution in the larger in challenging what might be construed by the smaller as its vital interests. The smaller country would be playing a riskier hand than the larger, for the latter would have an incentive and perhaps the capacity to eliminate the smaller one's nuclear force.   Even if the smaller country could damage—perhaps greatly—the larger, the larger could destroy the smaller. Therefore, there would remain large disparities in power and in advantages in bargaining.

Just as the standard " great power two-country model " leaves out essentials, there is a similar hazard in employing a simple model of a large country against a small one. This pair would not exist alone either. Small countries usually have ties with large ones; they may also be members of regional alliances. Introducing this possibility gives the small power alternatives which it would not possess by itself. For example, a second great power might give guarantees of support by nuclear weapons to a small country threatened by another great power. This, in fact, is common. Even powers of middle size armed with nuclear weapons find additional support useful. The United States provides such support to the two nuclear powers of middle size, Britain and France, against the Soviet Union. It is, however, increasingly questionable whether great powers will be as prepared in the future as in the past to provide such guarantees to countries in a nuclear crowd.

The Case of Two Lesser Powers

There could be many pairs of adversaries in which both had nuclear weapons; or in which one, as of a given date, would and the other would not; or in which neither possessed them, but both would be capable of acquiring them: *e.g.*, China and India, India and Pakistan, and Iran and Iraq, respectively. These binary situations would differ from the relationship of the United States and the Soviet Union in at least two important respects: their conflicts would most likely be direct and not expressed largely through support of third parties; and larger powers would be in the background and sometimes would be allied to the small ones.

Many of the relationships between small countries fit the standard " great power two-country model " better than does the relationship between the United States and the Soviet Union. Even so, its limitations remain substantial. Consider the widely held assumption that any use of nuclear weapons should be directed only against civilian populations. Pranger and Tahtinen, in their analysis of the threat of nuclear war in the Middle East, say that the use of nuclear weapons is much more likely between Israel and the Arab States than it is between the United States and the Soviet Union.[1] The United States and the Soviet Union, they hold, have developed a code of conduct for avoiding nuclear war whereas these adversaries in the Middle East have not. Through lack of experience in " living with the bomb " and as a consequence of deep hostilities, these governments would be much readier to use nuclear weapons, and probably to use them indiscriminately against civilian populations. This is a possibility which must be taken seriously, but there are also strong incentives working in such countries for restraint and discrimination in the use of such weapons.

The second difference between pairs of small and large countries is less speculative. Because a great power or one of middle size, which is the ally of a small one, can provide technology, arms and guarantees, the balance of power between two small countries could be rapidly altered by the transfer of weapons or of combat forces. This can be regarded as giving the small country limited " drawing rights " on additional resources. Alternatively, it can be treated as converting these into a conflict among more than two countries. Pursuing the latter approach, the larger allies of small countries might have several objectives: one would be to try to preserve a balance between the two small powers, for example, by sharing technology in order to reduce the vulnerability of forces; or by offering to replace the nuclear forces destroyed by its adversary if and only if the adversary used nuclear weapons first; or by using nuclear weapons directly against the adversary of an ally which launched an attack with or without the use of nuclear weapons. On the other hand, the large power might support its small ally with the aim of upsetting a local balance. The great powers have exhibited both types of conduct in their relationships with small powers.

An important parameter is the vulnerability of the nuclear weapons of a small country to attack by its immediate adversary. The vulnerability of

[1] Pranger, R. J. and Tahtinen, D. R., *Nuclear Threat in Middle East* (Washington D.C.: American Enterprise Institute for Public Policy Research, 1975).

nuclear weapons has been a central feature of the relationship between the United States and the Soviet Union for many years and this will be no less a concern in the relations of small countries. One side may think, perhaps correctly, that it can destroy the nuclear force of the other; where nuclear forces of both sides are vulnerable the fear of surprise attack could result in a small event giving rise to an attack with nuclear weapons. Indeed, the problem is more worrying than in the case of the great powers because of the capacity of powerful countries to transfer technology or information which could rapidly make opposing forces vulnerable.

In the case of some pairs of countries, one of the members will have nuclear weapons and the other will not—for a time. This is the situation today between India and Pakistan and between China and Taiwan—also perhaps, between Israel and Egypt or Syria. This relationship could exist, before 1990, between South and North Korea or South Africa and its neighbouring countries. The acquisition of nuclear weapons might be seen as deterring an attack with conventional weapons by an adversary without nuclear weapons; this might induce South Korea or Israel to produce nuclear weapons. The country with nuclear weapons might also try to coerce the one without such weapons, either through threats or the actual use of nuclear weapons. The attack with nuclear weapons on Japan by the United States in 1945 is the single historical example of such use.

The party without nuclear weapons in such a pair might accept the asymmetric relation or it might respond by seeking protection from an ally which already possessed nuclear weapons, or by acquiring its own nuclear weapons, or both. An effort to redress the balance by undertaking to produce nuclear weapons could lead to a period of high instability. For example, the Soviet Union apparently considered attacking the Chinese nuclear forces during the late 1960s during the period in which the Chinese were beginning to deploy nuclear weapons. Perhaps it did not do so because by then it seemed too late to prevent damage to the Soviet Union. Or perhaps there seemed to be no reliable way to keep China from rebuilding its nuclear stockpile short of permanently occupying the country, which would have been a formidable undertaking.

Among some pairs of small countries, neither will have nuclear weapons at the start. One or both might decide to acquire them out of fear of attack by conventional weapons or because one country is also a member of another pair in which it is subject to threats of attack by nuclear weapons. The fear that one might be second in the race can be a powerful motive. The Manhattan Project was stimulated in large part by a concern—unwarranted as it turned out—that Germany was vigorously working towards the bomb. The stability of the nuclear " overhang " would be endangered by awareness that the time for acquiring weapons once a decision to do so is made will be greatly shortened. " Races " to get nuclear weapons could be so short that it would be difficult to stop them.

The nuclear forces of the small powers will be small, probably without reliable systems to warn of enemy attack, and perhaps weak in safeguards against unauthorised actions by those in the chain of command. The forces of small countries are also likely to be specially prone to accidents and mistakes. The great powers possess the technological resources capable of reducing these hazards in the form of devices for locking, fusing, remotely controlling and releasing warheads. Making these available to small forces would reduce the dangers of unauthorised use. However, the safer nuclear weapons appear, the greater the incentive to acquire them in the first place.

Consequences of Nuclear Exchanges in a Nuclear Crowd

Much would depend on the setting in which the nuclear weapons were used, the actions of the great powers and, most importantly, on the immediate aftermath. For example, if the use followed the failure of a guarantee given by a great power or withdrawal of a guarantee, and if the consequences of the use were not disastrous, it might legitimate decisions to acquire these weapons and induce many other countries to do so. If the use resulted in large-scale damage to the country which used the nuclear weapon first, or if it drew very hostile reactions from one or more of the great powers, the example would be sobering.

The use of nuclear weapons by a small power might pose a threat so menacing to other countries that a larger power, or several together, might undertake action against it. Concern about the nuclear forces of small countries might cause the major powers to create conventional and nuclear weapons especially designed for use against these forces. Such weapons would be capable of being used with precision and with relatively small damage to population and to productive capacity. The purpose would be to reduce or eliminate the capacity of the small country to act mischievously, to discredit irresponsible leadership, and to strengthen factions favouring more moderate policies.

Effects on the General Stability of the International System

The proposition advanced by General Gallois and others that the wider dispersion of control over nuclear weapons would have a calming effect on international relations is doubtful. Not only is the period during which weapons are spreading to additional countries likely to be a period of exceptional danger, but the situation once the bombs have been made is also likely to be highly unstable. Beyond the 30 or more countries which may have ready access to fissionable material by 1990, there are well over 100 members of the United Nations which are in the queue. Moreover, there is a heightened possibility that groups within these countries might gain access to the weapons.

On the other hand, there seems little reason for the opposite conclusion that a world in which 20 or 30 countries possess nuclear weapons would inevitably produce a global holocaust, or that small wars with such weapons would be a common occurrence. Nevertheless, the threat of use of nuclear weapons will be more common and nuclear war will be more likely, including conflict between the great powers. Although we may assume that these powers will continue to work on the protection of their main nuclear forces in order to keep down the risk of surprise attack, uncertainty about commitments could lead to a sequence of actions and reactions which might culminate in a large-scale use of nuclear weapons.[2] In 1990, as in 1914, a series of moves in a complex situation could produce a world-wide catastrophe. To keep down the probability of such a disaster will be an important objective of the great powers. Such an objective would entail the protection of their own nuclear forces and the protection of those of allies. In order to do this, they will have to make their commitments to allies more explicit and more limited than in the past; for example, guarantees by great powers to small countries may increasingly be limited to deterring attack by an opposing great power using nuclear weapons.

Even without their actual military use, the dispersion of nuclear weapons will affect the world-wide distribution of power. Nonetheless, the pattern of the relationships among the countries of the world will probably not be wholly transformed. The British, French and Chinese nuclear weapons and the problems involving them have not substantially added to the effective power of these countries; even greater constraints will be faced by countries with smaller nuclear forces than these. (There is the real danger that nuclear weapons could come under the control of erratic or mad leaders. Such a development would very likely evoke a concerted counter-action by the threatened countries.)

Thirty years ago, Jacob Viner assumed that cities would necessarily be the targets of nuclear weapons because these weapons would remain expensive or scarce.[3] He also held the view, inconsistently, that many countries would make nuclear weapons in only a few years. Small countries with atomic bombs would become relatively more important because large ones would not have the prospect of nearly costless victories over them. He rejected as infeasible a world government as a response to the menace of widely distributed atomic bombs. But he considered the possibility that the dispersion of power might produce conditions favourable to the creation of a " balance of power " in which overwhelming concentrations of power could be avoided by the " timely negotiation of new alliances ".[4] This would restore balance temporarily lost. Great damage would be done by the use of these weapons in such a system. To maintain the system would require an extraordinary flexibility and speed in cancelling old alliances and adopting new ones. Moreover, power would probably not be widely enough distributed for an effective balance of power—or world government—to be feasible. He saw as the best, although slender, hope a " concert of great powers ", based on the United Nations Organisation together with the unremitting practice of mutually conciliatory diplomacy.

Thirty years later, some of Viner's predictions have turned out to be false and others, in consequence, are untested. The prospect of world government seems at least as remote as it did to Viner in 1945, and the

[2] For an analysis of this and other possible sources of instability in a nuclear world, see Hoffman, Stanley, in Buchan, Alastair (ed.), *A World of Nuclear Power?* (Englewood Cliffs, N.J.: Prentice-Hall, 1966).

[3] Viner, Jacob, " Implications of the Atomic Bomb for International Relations ", paper presented to the American Philosophical Society, 1945, reprinted in his *Essays in International Economics* (Glencoe, Illinois: The Free Press, 1951), p. 300.

[4] *Ibid.*, p. 307.

United Nations has hardly been a successful mechanism for the maintenance of peace by the great powers. It is doubtful that the fundamental differences which separate the great powers will be overcome by the shared sense of danger from the small ones sufficiently to enforce concerted action upon them. At least, this does not seem likely to occur soon. The alternative of a highly decentralised system now seems unlikely to emerge; the "levelling" tendency of nuclear weapons is not powerful enough. One can only conjecture whether the existence of a system of widely dispersed power could provide a stable international order. Equilibrating alliances might be formed and reformed. Although there could be a long-term interest to participate in collective arrangements for security, there might be thought to be a short-term advantage to abstain from them. In short, there does not seem to be a good basis to hope for benign consequences from the dispersion of nuclear weapons.

The Future of Guarantees in a Nuclear Crowd

Let us imagine two very different patterns in the world. In one, all countries would be members of tightly linked systems of collective security, perhaps some of them highly hegemonic in character. Athough not all interests would be shared by the members of each alliance, there would be much confidence among them in the collective arrangements as they would bear on essential issues of security. It would be expected that each alliance would persist. The likelihood of independent military action by individual countries would be low. In the other pattern, there would be no fixed alliances. Each country would perceive itself as isolated from others on essential matters of security except for temporary coalitions. Countries would act independently.

The former would be a world of relatively little uncertainty in which there would be little demand for nuclear weapons and much willingness to give guarantees to offset threats; wars among small countries—*e.g.*, Arab-Israeli wars—would be likely to be quickly stopped by large ones. The second pattern would exhibit the opposite tendencies. Although in the first pattern, conflict among the blocs would be possible, wars would be more likely in the second. The conflicts which might emerge would be limited to small powers but they might nevertheless be catastrophic to the combatant countries.

Neither pattern corresponds to the world as it has existed in the past or is likely to in the future. But the world has moved during the past 20 years from one which was largely bi-polar and predictable to one which is much more multi-polar and less predictable. Even in the high period of bi-polarity, some countries were outside the main systems of alliance or on their fringes: for example, South Africa, the countries of Black Africa (after decolonisation), and the countries of South and Southeast Asia. But almost all the economically advanced countries were, more or less firmly, members of alliances. This is the main reason why most of the advanced countries are now in the "overhang" rather than in the category of those possessing nuclear weapons. This is most obviously true of Western Europe. As for Eastern Europe, the Soviet Union provides a combination of protection of and coercion of its allies. The question of the production of nuclear weapons does not even arise in the German Democratic Republic or in Poland. The United States has no such power over its allies, and many of them could rapidly acquire weapons today.

The non-communist alliances centred on the United States have become eroded. This erosion is a product of the failure of the United States in Southeast Asia, isolationist tendencies within the United States, and a reduced sense of danger from the Soviet Union, which is accompanied by the increased military power of the Soviet Union. It is significant, however, that none of the countries which today seem most likely to acquire nuclear weapons by 1985 are members of NATO, despite the fact that they, with Japan, are the most technologically advanced countries. NATO, together with European institutions, provides bonds which still give a very substantial degree of security to these countries. This is also true for countries such as Sweden and Switzerland which are outside the alliance but which benefit from its existence. The arrangements for security between the United States and Japan perform the same function for Japan.

These alliances not only offer protection from adversaries, they restrain competition among their members. The integration of the Federal German Republic into the European Economic Community and its participation in NATO has played a crucial role in preventing pressure by West German and other Western European countries for the acquisition of nuclear weapons. The alliance has also contributed towards moderating the competition between the United Kingdom and French nuclear weapons programmes, in part perhaps because the British programme has been linked to the United States.

The countries most likely to acquire weapons are the " non-aligned " and marginally " aligned " countries and those which feel threatened and fear abandonment. These include the Republic of Korea, Israel, Taiwan, Argentina, Brazil, Pakistan, Iran, South Africa and India. South Korean confidence in the United States has been affected by the collapse in Southeast Asia. Israel, which recognises its great dependence on the United States, nevertheless fears that it might be coerced by it into making fatal concessions or that it might be left in the lurch in a crisis. Taiwan fears the breaking of ties with the United States. Argentina and Brazil are rivals and are members of a decreasingly effective system of collective security in the Western hemisphere. The United States is providing arms to and has been supporting Pakistan and Iran, but their ties with the United States are hardly close. South Africa has been isolated and India is officially " non-aligned ".

Proliferation will affect both the demand and the supply of guarantees through alliances. The demand for support by alliances will increase among those countries threatened by rivals which have acquired nuclear weapons. Pakistan is in this situation today. It seeks support from other countries in the region, *e.g.*, from China, Iran and the Arab countries. Countries which feel threatened are likely to seek the support of larger powers; for Pakistan this is largely the support of China and the United States.

It is doubtful that the desired alliances will be forthcoming. The great powers might argue that countries with nuclear weapons no longer need guarantees, or that guarantees to such countries in a region with already established nuclear powers would be too dangerous to themselves.

The reluctance of the great powers to provide the guarantees entailed in alliances could create instability. A reduced willingness to undertake such obligations would increase the incentives of small countries to acquire nuclear weapons, which in turn could strengthen the large power's inclination to pull back. The geographical isolation of the United States and the nature of its ties with the rest of the world make it more likely that it might be caught up in such a process than might the Soviet Union. The proximity of the Soviet Union to many other countries, together with the hegemonic nature of its domain, makes it much less likely to withdraw. It might, however, become more restrained about taking on new and distant commitments.

Finally, another alternative for powers of small and middle size is neutrality and accommodation. This would mean the renunciation of nuclear weapons so as not to appear threatening. It would not be surprising if a number of countries in the " overhanging " category choose this path.

" Nuclear Linkages "

Decisions in one country to produce nuclear weapons will cause reaction in others, near and remote. One pattern of a branching process in the acquisition of nuclear weapons would be the following: Country B decides to obtain weapons because Country A has them or is getting them. But B also is a rival of, or is actually hostile to country C, which now has more reason to acquire nuclear weapons. Consider Asia. China acquired nuclear weapons for several reasons, including its rivalry with the Soviet Union. This act had a decisive effect on India which, after a lag of 10 years, exploded a nuclear bomb. Another of its adversaries, Pakistan, now feels threatened and begins a programme to produce nuclear weapons. Iran, which in any case is worried about dominance by the Soviet Union, regards India as a rival. These factors might lead to an Iranian decision to make weapons. At this stage, a number of neighbouring Arab countries, which have in any case been concerned about a possible Israeli nuclear force, fear the consequences of an Iran strong in both non-nuclear and nuclear weapons. This might cause them to create the preconditions for obtaining nuclear weapons. Subsequently, most or all of these countries might take the step of producing nuclear weapons.

This analysis is oversimplified. For example, it leaves out the influence of alliances. But it also suggests that the existence of multiple hostilities and rivalries will produce a sequential branching process. The situation will be aggravated when there are many countries in the " overhanging " category.

Are there likely to be " stopping points " which could halt such a process—at least for a considerable period? Let us consider Western Germany and Japan. Many analysts predicted in the 1950s and 1960s that it was inevitable that powers of the middle rank would soon acquire nuclear weapons and therefore that Western German and Japanese tests of nuclear weapons would follow the British, French and Chinese by about 1970 or 1975. Within Europe, it was anticipated by these analysts that Sweden, Switzerland, Italy and others would follow in turn or more or less simultaneously with Germany. This has not happened nor is it imminent. The reasons are clear. The Federal German Republic was at first inhibited by its concern about internal stability, and it has been constantly concerned about the fears and consequent reactions which a programme for producing nuclear weapons would call forth both in Western and Eastern Europe. It has also placed a good deal of confidence in its alliance with the United States. Japan, less exposed to attack than Germany, has sought to excel in the economic rather than the military realm. Japan too benefits from protection by the United States. The " stopping points " experienced by these countries have reduced the pressure on others to go ahead with the production of nuclear weapons.

Similar " stopping points " are not evident elsewhere in the world—except for the countries of the Warsaw alliance in which the Soviet Union effectively acts as the " stopper ". " Stopping points " require a stable set of relationships within a region and with outside powers. Such stability exists mainly in Western and Eastern Europe. Elsewhere, relationships are more fluid.

The Weakening of the Nuclear Taboo

The Indian explosion of a nuclear bomb seems to have had an effect on distant countries; for example, Argentina and Brazil. The acquisition of such weapons by additional powers could have an effect throughout the world in weakening the inhibition against the acquisition of nuclear weapons, even on such industrial powers as Western Germany and Japan. It would weaken the widely shared taboo on weapons.

This weakening could have a serious effect on the internal debates about the desirability of acquiring nuclear weapons. In such debates, not only are practical considerations important, for example, anticipation of the responses of the United States or the Soviet Union, but also beliefs about the legitimacy and morality of nuclear weapons. The acknowledgement of a hierarchy of states according to wealth and power has been a significant factor too but this is now being increasingly challenged. It has been regarded as one thing for the great powers to have these weapons or even for other powers possessing a high level of technology such as Britain or France. But for poor countries such as Pakistan or Korea to acquire nuclear weapons has been thought to be a different matter. The economically advanced countries are likely to have their policies of abstention from the acquisition of nuclear weapons strained regardless of whether they are directly threatened. Although the factors which have prevented Germany and Japan from acquiring nuclear weapons remain powerful, their spread to a large number of smaller countries would greatly weaken the abstension of these middle powers. (There might also be more direct regional repercussions on these two countries. Japan, for example, would be profoundly affected by a nuclear programme in the Republic of Korea and Western Germany could be affected by the production of nuclear weapons in Spain or Yugoslavia.)

The acquisition of nuclear weapons by the Federal German Republic would probably have serious repercussions in Europe. As for Japan, if the Japanese decide to acquire nuclear weapons they will have a sizeable programme, and Japan, armed with such weapons, might appear threatening to many other countries. Such developments might be accompanied by far-reaching changes in Japanese politics.

These reactions are not the only possibilities. Either Western Germany or Japan might turn towards neutrality and passivity.

The Proliferation of Sources of Supply

The market for nuclear technology has changed greatly in the past 20 years. Until recently, only the United States, the United Kingdom, Canada and the Soviet Union were significant producers of nuclear electrical power stations; the United States dominated the export market with Canada in a secondary role. Now France and Germany play a major role in this market and Japan and Sweden could soon do so. Reactors derived from the technology developed by Westinghouse are sold by Kraftwerk to Brazil. Successful South African development of the capacity for enrichment of nuclear fuel, helped by German technology, would enable it to sell not only enriched fuel but also plants for enrichment. Western Germany has already contracted to sell Brazil the technology of jet-nozzle enrichment. Western Germany, or its partners in Urenco might, in the future, sell the more advanced centrifuge technology now under development to other customers. India might—though there is little evidence of this yet—become a supplier of technology more directly related to nuclear weapons, such as the technology for the reprocessing of spent fuel.

This increase in the number of supplying countries will greatly reduce the effectiveness of unilateral decisions by any one country, or a small group of them, to deny access to nuclear technology to others. Virtually any advanced industrial country of middle size, to say nothing of the larger ones, could, in time, become a supplier of nuclear components, and more than a few might be able to provide entire plants. On the other hand, the number of countries which will be able to supply certain very important types of equipment needed for the manufacture of nuclear weapons, such as the technology of enrichment or reprocessing, will remain small for the next decade. The adoption of strict limitations among the exporters of nuclear equipment could make a big difference to the diffusion of such important technology, but this close cooperation is only now evolving.

A specific problem related to nuclear weapons is the transfer of separated plutonium and of highly enriched uranium from one country to another. Significant quantities have been sent by the United States and other supplying countries for research, testing nuclear materials and for ship-reactors. Even with safeguards, this is a dangerous precedent. Other suppliers of such material among the countries which are expected to have at least 250 kilograms of fissile material by 1985, might have less concern than does the United States about what is done with their material; or they may simply be unable to apply sanctions if agreements are violated. Furthermore, the growth in the number of potential suppliers is large enough for the existence of a market in fissile material useable for nuclear weapons. The possibility of a market in fabricated weapons cannot be excluded.

The Continuing Competition in the Technology of Nuclear Weapons

The initial testing and construction of a few nuclear weapons is only one stage in a process. Qualitative changes in technology will continue to take place, and this will require continuing changes in nuclear forces just as it does in conventional forces. The competition will include development of thermonuclear devices, more advanced vehicles for the delivery of nuclear weapons, basing methods, warning systems and surveillance and command and control systems. If a country does not compete in these respects, its forces will become increasingly vulnerable and unreliable in performance. One should not assume, however, that a series of " uncontrolled " nuclear arms races will be set off by these pressures. Limitations of resources and the existence of competing domestic and military objectives will limit expenditures on nuclear forces. But the very existence of such forces becomes of great symbolic importance—and dangerous—to their possessors.

The situation which will be faced by small powers with nuclear weapons is suggested by an examination of the resources devoted to nuclear forces today. Let us consider the three levels of expenditures for research in the production of nuclear weapons, specialised delivery systems for nuclear weapons, strategic intelligence, and command and control systems, and, for the United States and the Soviet Union, air defence of their national territories.

Great Power:	$10 - 20 billion annually (United States and Soviet Union)
Middle-sized Power:	$ 1 - 2 billion annually (United Kingdom and France)
Small Power:	$ ·1 - ·4 billion annually (Israel? India?)

According to the theory of minimum deterrence, the effective power produced by widely different levels of expenditure should be similar. However, governments can apparently discern the difference between the military capacities of forces which differ by factors of 10 to a 100. Great Britain and France have not produced nuclear forces which appear to anyone to be comparable to those produced by the Soviet Union through much greater expenditures.

How do these three scales of expenditure on nuclear forces compare with the resources likely to be available for the acquisition of nuclear weapons by the countries which might produce nuclear weapons in the near future? In making such an estimate one should recognise that only under circumstances of great stress will a country devote a large share of its resources to military matters. Israel is unique in devoting about 30 per cent. of its gross national product to military purposes; the Soviet

Union probably devotes around 15 per cent.; most countries allocate around 5 per cent. or less to defence. Most of the existing nuclear powers devote about 1 per cent. of their gross national products to their nuclear forces. This is a rough estimate; it is especially difficult to make for the Peoples' Republic of China. The Soviet Union is exceptional in devoting about 2–4 per cent. of its gross national product to these forces.

Among the countries expected to have enough fissile material for three or more bombs in 1985, only Japan and Western Germany of the powers of middle size could easily allocate resources to military nuclear programmes. If they desired to do so, each of these countries could proceed on the scale at the current level of the over $10 billion annually which is spent at present in the great powers. Among the other countries in the " overhanging " category, probably only Iran—among those which seem to have an incentive to acquire nuclear weapons—could come close to supporting a programme of a power of the middle rank. For the others, it would be difficult to have any programme beyond that of a small power, although an acute sense of danger might stimulate an attempt to do more.

Even though most new nuclear powers are likely to devote two orders of magnitude fewer resources to nuclear military programmes than the great powers and one order of magnitude less than the powers of middle rank, their nuclear forces will enable them to cause a great deal of trouble. If these forces are not to be very dangerous to their possessors, they will require close attention to the safety and protection of the weapons and control, security against internal dissidents, and careful planning of their possible use. It might be possible for small countries to satisfy these requirements and to be able to use their weapons without suicidal consequences, but this will not be easy.

The contribution to stability which great powers can make in a world with small nuclear powers raises questions about the objective which is embedded in the currently dominant theory of arms control and which is contained in the Non-Proliferation Treaty: namely, that the great powers should sharply reduce and ultimately eliminate their stocks of nuclear weapons. To narrow the difference between the large and small nuclear powers will not increase stability; it will not be reassuring to those smaller countries which depend on nuclear guarantees from large ones. The preservation of the difference between the large and the small nuclear powers may become important as the " nuclear crowd " grows. It is certainly desirable that nuclear weapons be one day entirely eliminated, but the reduction of the superiority of the great powers will not contribute greatly towards that end. And it could hurt.

Prospects for Further Increases in the Size of the " Crowd "

Nuclear spread will not have completed its course by 1985 or 1990. By then about one fourth of the more than 150 members of the United Nations will be in the category of those capable of producing three weapons or more. There will remain many potential entrants during the following decades. The extent to which the remaining countries move into the "overhanging" category or actually produce weapons and the rate at which they do so will depend on six factors. These are: whether or not nuclear electrical power becomes economic in small and poor countries; what happens to the countries which have acquired weapons; the nature of the market in nuclear technology—including possible markets in separated plutonium and highly enriched uranium or even nuclear weapons; the degree of security provided by alliances; the nature of the rivalries in various regions and the extent to which local agreements, formal or informal, might limit the spread of nuclear weapons; and finally the extent to which the "nuclear taboo" is broken. Prediction is even more difficult here. We should not assume, however, that a world of *omnium contra omnes* is inevitable. There will be increasingly powerful incentives to keep it from emerging.

What will Proliferation do to Doctrines of Nuclear Warfare?

The doctrine that nuclear weapons should be aimed against civil populations has dominated most of the academic literature in the United States, some of the rhetoric of government, but less of its planning of operations. According to this doctrine, the task of a nuclear force is in principle simple: it is to survive a well-executed attack by an adversary —which in a world of proliferation might be a great power—reliably receive the order to launch, penetrate to enemy centres of population and to destroy them. This task is less simple in execution. The United States has not been confident of its ability to manage it without a continuing effort. (Ironically, this doctrine has in the past been advocated by some Americans who, following General Gallois, proposed that the United States should provide small nuclear forces to its European allies, who could aim them at Soviet cities as a means of reinforcing European security and as an alternative to the provision of "nuclear guarantees" by the United States.)

This doctrine is likely to be espoused by some leaders of small nuclear countries and some may even act on this theory. It is possible, however, that the governments which acquire nuclear weapons may behave more responsibly than the United States and the Soviet Union have occasionally appeared to intend. For one thing, irrational policies are most readily supportable if they are not going to be acted on; many small countries which are hostile to each other confront each other directly. Especially where situations in which nuclear weapons might be involved are two-sided, which means in most cases, the incentive to develop alternative plans for the limited use of nuclear weapons not aimed at civilian populations will at least on the average be great; catastrophic attacks on civilian populations are not to be ruled out, but they are unlikely to occur often. The small powers have good reasons to use nuclear weapons, if they use them at all, in very restrained ways—except perhaps as a last desperate resort.

Alternatives in the use of nuclear weapons by Israel: The problems faced by small countries are illustrated by the case of Israel. The possible existence of Israeli nuclear weapons did not prevent the Arab attack in 1973. In the future, how might Israel use such weapons? If used at an early stage in the conflict against the Arab armies, there would be an uncertain tactical effect unless more than a few weapons were used. Furthermore such use or even the direct threat of such use would give a licence to the Soviet Union to provide nuclear weapons to the Arabs—if they did not already have their own. There might be a direct attack with nuclear weapons on Israel by the Soviet Union. As for the United States, it would no doubt warn the Israelis not to use the bomb, warn the Soviets not to give them to the Arabs, and warn the Soviets and the Arabs not to use them against Israel. Threats to use nuclear weapons against Cairo or Damascus or their actual use by Israel would produce very strong reactions from foes and friends.

In short, if the Israelis were to threaten use or actually to use nuclear weapons too early, before their backs were to the wall, they would lose the support of the United States and they would give a licence to the Arabs or the Soviet Union to retaliate; if they were to use them late, that might not prevent the destruction of Israel—indeed, it might assure it. Only limited, clearly defensive use would reduce the likelihood of a devastating response to a possibly tolerable probability. But by how much? Even limited defensive use could provide a plausible excuse for the use of nuclear weapons against Israel as a response.

Might the Israelis try to deter such a Soviet response by trying to acquire the capacity to attack Soviet military forces on the territory of the Soviet Union? Such a force would probably be vulnerable on the ground, and it would also be vulnerable in attempting to penetrate the air defences of the Soviet Union. Israel might try to deal with its vulnerability on the ground by basing nuclear weapons at sea. (Israel has two submarines and three on order. These could be equipped with cruise missiles such as might be developed by Israel over the next decade.) A submarine force would have to withstand intensive search by the technique of anti-submarine warfare. An attack on Soviet forces in the region, as on Soviet ships in the Mediterranean, would be painful for the Soviet Union, but losses to its fleet would be regarded as a justification for the elimination of Israeli military power and to provide a powerful demonstration of the power of the Soviet Union; it might be willing to make such a trade. If Israel were to build a missile force able to reach cities in the Soviet Union, even Moscow, such a force might give the Soviets pause. It might deter a direct attack by the Soviet Union on Israel. On the other hand, it would give the Soviet Union an even stronger incentive to obliterate Israeli power. It might view aiding the Arabs to acquire nuclear weapons as the least dangerous way to deal with this threat. The Israelis are not likely to attack Moscow with nuclear weapons in reaction to Soviet military assistance—even including the giving of nuclear weapons—to the Arabs.

The essential point is that Israel is a country likely to be highly sensitive to the need to avoid indiscriminate attacks on civilian populations. At best, it appears that Israeli nuclear weapons could be reserved for deterring initial use of nuclear weapons by enemies, or for selective use if Israeli armies were on the verge of defeat under circumstances in which the survival of the country was at stake.

What are the prospects of Arab threats of the " first use " of nuclear weapons or of actual attack? There is much asymmetry in area and population between the two sides. Could some Arab leaders at some point decide that the elimination of Israel was worth the cost of Israeli retaliation? More likely is the possibility that instability in an Arab country possessing nuclear weapons could cause some of these weapons to be transferred from prudent hands to those eager to use such powerful weapons. Nor should we overlook the reports that President Khadafi in Libya, a leader not noted for restraint, has an offer outstanding to purchase nuclear weapons.

The defence of Israel will continue to depend largely on its own non-nuclear forces backed up by the power of the United States.

The alternatives before the Republic of Korea in the use of nuclear weapons: Korea, in the opinion of many informed military experts, is capable of defending itself against a North Korean attack with conventional forces. This opinion is close to being concurred in by President Park who has been recently quoted as saying that Korea will be able to defend itself against the North by the end of a five-year programme of armament. On the other hand, it could not withstand an attack in which China or the Soviet Union participated. Korea will continue to need guarantees of support by nuclear weapons against attack by nuclear weapons, or it will need the guarantee of support by conventional or nuclear weapons against a larger attack by conventional weapons than could be made by North Korea alone.

A Korean programme for the production of nuclear weapons then would be motivated largely by uncertainty about the continuation of American support. If the commitment of the United States ends, how might Korea use nuclear weapons? Retaliation on Pyongyang in response to a non-nuclear attack from the North, would, as in the case of Israel, invite and give a licence to the Soviet Union or China to aid North Korea in retaliating against Seoul or against its military forces. Attack by nuclear weapons along the demilitarised zone appears to be tactically unnecessary, given the defensibility of the narrow invasion routes with conventional weapons. Such use would also provide an excuse for retaliation by North Korea or its allies by means of nuclear weapons.

Concern expressed by officials of the Republic of Korea about the threat of attack on Seoul by " frog " missiles without nuclear warheads provides another reason, or rather rationalisation, for a programme of production of nuclear weapons. Korean fear that an attack on Seoul could create panic and chaos does not seem to be warranted by a quantitative assessment of the damage which could be done. A " frog " missile aimed at the parts of Seoul with high density of population would probably kill between 10 and 100 persons. North Korea would need very large forces to do large-scale damage to Seoul with missiles or aircraft carrying high explosives. In contrast, a 20 kiloton nuclear weapon on Pyongyang might kill about 160,000 persons and a one megaton weapon on Seoul in response would kill about 2·7 million.[5] The Korean officials could be more reassuring to their people on the relative unimportance of the dangers which could be done by non-nuclear attack and they would be able to reduce the likelihood of panic if they were clearer on these facts themselves.

The alternative for South Korea of assuring an effective defence with conventional weapons along the demilitarised zone, and of counting on the United States to keep the Soviet Union and China out of active military involvement, is not certain to be effective. It is, however, much less dangerous than the alternative of nuclear weapons.

The alternative of adopting a " no-attack-on-population" doctrine: We should consider adopting and advocating a rule which would outlaw attack on cities by nuclear or other weapons. This rule has been proposed by the International Commission of the Red Cross for the Middle East and has been favourably received by Prime Minister Rabin. The adoption of such a rule might help to calm fears in Israel and Egypt that attacks on civilian population are to be expected in the next war. Given the frequency of wars in the Middle East, this is by no means a trivial worry. Moreover, it might make a difference in actual conduct in a conflict. Many wars have begun with countries which were restrained from attacking non-combatants but which succumbed to the opportunities and temptations during the course of the war. There is some basis in history for hoping that future conflicts in the Middle East could be stopped before the process had progressed so far.

This proposal is related to the recent dispute over supply by the United States of Pershing II missiles to Israel. The Israelis have argued that they need Pershing missiles to offset the Arab possession of Soviet Scud missiles which are able to reach the whole of Israel from places well back in Arab territory. They have also argued that they are needed for use in combat against military targets since the Pershing II will be much more accurate than the Pershing I. A third Israeli argument is the need to be able to reach the more distant Arab countries in order to try to deter them from joining in a future war. The Israelis are concerned that Scud missiles could be launched randomly against Tel Aviv, Haifa

---

[5] Some current estimates of the increase in deaths which would be brought about by substituting nuclear for conventional weapons in wars outside the West have understated the difference this change would make. Like the very early estimates of P. M. S. Blackett, the nuclear calculations may be based on a gross analogy with Nagasaki rather than on the particular terrain and patterns of settlement of the cities in question. Although the number of deaths may seem large compared with those at Nagasaki, Nagasaki is a poor example of the destructive power of nuclear weapons. At Nagasaki, the bomb missed the aiming point by nearly two miles and the hilly terrain shielded parts of the city from the destructive effects of the bomb. The larger number of deaths which might occur in Seoul is only partly attributable to the use of a nuclear weapon of higher yield. A well-placed 20 kiloton nuclear weapon on Seoul might kill over 350,000 persons. These figures are conservative since the effects of thermal radiation are ignored. The calculations of the effects of high explosive warheads in Korea are based on a detailed settlement map of Seoul and of small population units called *dongs*. For the 20 kiloton and the one megaton estimates, average densities were used and checked for the approximate correspondence to the details of the pattern of settlement.

and Jerusalem during a conflict. Within one week, this could produce, unless countered, over 1,000 deaths and a large number of casualties. It could, in the view of some Israelis, be highly disruptive to Israeli morale and to their effectiveness in war. This argument supports the position that the appropriate deterrent is an Israeli missile capability to carry out a similar campaign against the Arabs. In this view, missiles are needed rather than aircraft because of their especially terrifying character.

It is not easy to assess the arguments on the psychological efficacy of one type of weapon as compared with another. But what does seem clear is that the Pershing II, even though much more accurate than its predecessor, is likely at best to be effective against only a few types of targets; it will probably be looked upon as principally a carrier of nuclear weapons, not as a vehicle which would be used for conventional warheads; and a policy of buying missiles which appear to be mainly useful as carriers of weapons against Arab centres of civilian population—especially given hints about Israeli possession of nuclear weapons—seems imprudent and inconsistent with the goal of calming concerns about and reducing the likelihood of actual attacks on civilian populations. The alternative of focusing attention on avoiding attacks on civilians deserves a good deal more attention than it has received.

New Threats to the Forces and Territory of the United States

Americans who belittle the importance of proliferation, or even suggest that some proliferation might help the United States, tend to ignore the new threats to American forces abroad and to the territory of the United States which could result. These threats are of several kinds: the United States could be subjected to direct threats from the small countries, even former friends.[6] The United States could come under nuclear attack as the result of circumstances it could not control; for example, through an ally taking actions which result in a nuclear response against American military forces. An example would be an attack by a South Korean force equipped with nuclear weapons against the North leading to a response by the Soviet Union against South Korea. Such retaliation could hardly not affect the American forces in Korea, assuming that they were still stationed there.

One reaction by the United States to the possession of nuclear weapons. by especially dangerous governments could be an attempt to eliminate such a force. This is an action which the Soviet Union appears to have considered with respect to China. But more likely is the cessation of American commitments to countries which acquire nuclear weapons. This, however, would increase the incentives to acquire nuclear weapons and to increase investments in existing nuclear forces. This, in turn, would create an increased capacity to attack the forces and even the territory of the United States.

---

6 French assertion of a policy of defence à tout azimut in the 1960s should have given pause to American enthusiasts for the dispersion of nuclear weapons.

By the late 1980s, several countries might be capable of delivering nuclear weapons against the continental United States. As a result, it might find it desirable to allocate offensive forces to deal with the nuclear forces of such countries and to build up air defence and civil defence, and possibly anti-ballistic missiles, to cope with these threats.

The problems which might be created for the United States and its military forces vary a good deal among regions.

Northeast Asia: A threat to American forces could occur from the production of nuclear weapons by South Korea, Taiwan or Japan, with all of which the United States has security agreements. For example, the use of Republic of Korea nuclear weapons could result in nuclear retaliation by nuclear weapons against American forces in Korea or elsewhere in the region by the Soviet Union or the Peoples' Republic of China. Taiwanese nuclear weapons would present a similar risk to American forces in Taiwan if they continued to be stationed there—a prospect which is becoming increasingly unlikely. Nuclear weapons in the possession of a hostile Japan would present a serious threat to American forces given the large economic and technological resources available to Japan. For example, Japan is developing the capacity, through its space programme, to build missiles able to reach the continental United States, although any such missile force might be vulnerable to American counteraction.[7] Fortunately Japan is most unlikely to change its present doctrine of self-defence which eschews nuclear weapons. Thus, American forces in East Asia and the Pacific could be exposed to attack with nuclear weapons by five countries in the region by 1985 if all of those in the " overhanging " category acquired weapons, *i.e.*, by the Soviet Union, China, Japan, the Republic of Korea, Taiwan. Of these, three—the Soviet Union, China, Japan—could have the capacity to reach the continental United States.

South and Southeast Asia and Australia: The prospective possessors of nuclear weapons by 1990 in this region are China, India, Pakistan and Australia—and perhaps, later, the Philippines and Indonesia. Either India or Pakistan could threaten American forces and bases in the Indian Ocean and Persian Gulf areas, *e.g.*, the American base being built at Diego Garcia. India by 1990 might have a capacity to build ballistic missiles able to reach the United States. The Philippines or Indonesia could threaten any American forces which might be in Southeast Asia.

Middle East and North Africa: The most probable nuclear powers in this region by the late 1980s, Israel, Iran and Egypt, could threaten American forces in the Indian Ocean, the Persian Gulf and the Mediterranean with

---

[7] Any country with long-range civilian jet aircraft has the ability to carry nuclear weapons over intercontinental distances. Nuclear weapons might also be carried on ships, merchant or naval. Having the ability to carry nuclear weapons a long distance does not mean that such a force would be well protected or able to penetrate reliably to target. The long-range delivery capacities discussed here refer to ballistic and cruise missiles and military aircraft.

the aircraft and the missile systems which these countries will possess.[8] They are unlikely to be able to attack the continental United States.

Sub-Saharan Africa: The only prospective possessor of nuclear weapons in this region by 1985 is South Africa. The United States has no forces in the region nor is South Africa likely to expend resources on intercontinental systems.

Europe: At least 20 countries in Europe could have nuclear weapons by 1985. In addition to the Soviet Union, United Kingdom and France, 17 countries will have accumulated substantial stocks of fissile material by that time: Austria, Belgium, Bulgaria, Czechoslovakia, Denmark, Finland, the Federal German Republic and the German Democratic Republic, Hungary, Italy, the Netherlands, Portugal, Rumania, Spain, Sweden, Switzerland and Yugoslavia. (The Soviet Union will try, presumably successfully, to prevent the countries in Eastern Europe from acquiring weapons.) Any of them could pose a threat, directly or indirectly, to American forces in Europe, the Mediterranean or the North Atlantic. Among these countries, the only potentially new nuclear power which has the resources to build a substantial force capable of reaching North America is the Federal German Republic.

Western Hemisphere: The United States does not have forces deployed in the neighbourhood of Argentina and Brazil, which are the two most probable possessors of nuclear weapons in this region. Nor are they likely to acquire long-range delivery systems beyond the commercial jet aircraft which they own. Canada has long had the capacity to build nuclear weapons, but there is no evidence that it will do so. Mexico is developing the capacity to produce nuclear electrical power and probably could acquire nuclear weapons by the late 1980s, but there is no evidence that it intends to follow this path.

The Prospects for Defence by Conventional Weapons as a Factor Affecting Proliferation

Some of the countries which will be in the " overhang " category have defence against attack by conventional weapons as a principal concern. Some of these countries have terrain highly suited for such defence, *e.g.*, Korea, and—for all of them—the alternative of employing advanced non-nuclear technology could greatly increase their defensive capacity.

Some leaders of countries which feel threatened have made explicit the relation between defence by other than nuclear weapons and the acquisition of nuclear weapons. For example, when Mr. Bhutto, Prime Minister of Pakistan, said that if the United States did not provide conventional arms to Pakistan and the disparity between India and Pakistan reached the stage where it threatened the stability of South Asia, Pakistan might be forced

---

[8] The possibility that Libya or other countries in the Middle East could acquire some weapons by 1985 through purchase cannot be excluded.

into the production of nuclear weapons. Bhutto's position is likely to be emulated by other heads of government.

In short, making modern non-nuclear weapons available to countries concerned with the danger of attack by weapons other than nuclear ones might, as part of a larger policy, contribute significantly to weakening the drive in some countries to obtain nuclear weapons.

Nuclear Threats by Extra-Governmental Groups

The countries with nuclear weapons are among the more stable countries. Many of the prospective possessors in the next decade have been less stable. Korea, Pakistan, Egypt, Argentina, Brazil and Iran are not among the most unstable states, but there is a substantial history of attempted and successful *coups d'état* in each.

In countries with nuclear weapons, plotters aiming to seize power will make gaining control of them a major task. Although their use in *coups* and armed uprisings seems very unlikely, capturing the nuclear weapons will be high on the agenda of the putschists. For this reason, heads of government are likely to pay particular attention to the reliability of the commanders in charge of nuclear units. Concern about this problem may produce a conflict between those guarding weapons from domestic enemies and those protecting them from foreign ones. The former suggests that they should be kept in the local garrison under reliable control; the latter suggests that they should be widely dispersed and concealed. One way in which these two conflicting objectives might be partially reconciled is for weapons to be provided with permissive action links with the keys held by the head of the government and perhaps a trusted aide or two. (The question then will be whether the holders of the keys will be more vulnerable than the weapons.)

Use in domestic conflicts, although unlikely, is not excluded. Some civil wars have been marked by extraordinary bitterness and brutality. A government might find the temptation to deliver a crushing blow against rebel strongholds with nuclear weapons overwhelming; rebels, if they gained control of some weapons, might also find the demonstration of their power irresistible. However, rational participants would put political criteria uppermost and targets would probably be selected which would avoid great civil damage but would have great political effect. Moreover, those which have geographically identifiable bases of support share an important vulnerability with governments—they have bases against which retaliatory action can be taken. Most importantly, governments have a powerful incentive to retain a monopoly on this form of power and to take measures to minimise the likelihood of losing control.

Conclusions

The foregoing analysis has presented the need for significant changes in the international rules for nuclear power. The evidence is strong that the benefits of nuclear power can be retained along with restraints to lessen its dangers. Arguments that the world will soon run out of uranium and that plutonium must be used as a fuel have little basis in fact. Nor is it necessary to use highly enriched uranium, another explosive material, as a fuel. Looking ahead, several subjects need additional work: one is to try to achieve greater recognition throughout the world of the common danger which will result from wider access to nuclear explosive materials. This perception is not widely shared today. Another is to expand greatly research and development on alternative and safer forms of nuclear power than the plutonium breeder or other technologies which require the circulation of explosive materials. Not only new technologies but new institutional arrangements will also have to be devised. Restraints on the spread of dangerous technologies, although necessary, are not sufficient. Economic, political and security incentives also must play a major role. Clearly, new international rules which might be widely accepted will have to take account of the diverse situation of countries with respect to access to fuels, widely differing levels of technology and different perceptions of dangers to security. Devising non-discriminatory rules which will make the benefits of nuclear power widely available and limit its dangers, while at the same time meet diverse national needs, is a formidable task. However, the consequences of the spread of nuclear explosives warrant making a great and sustained effort to limit the worst risks.

APPENDIX A:
World Nuclear Facilities

1. Civilian Nuclear Facilities and Their Military Potential

A. Introduction

THE primary technical obstacle to obtaining nuclear weapons has been the inaccessibility of high concentrations of one or another of the fissile isotopes uranium-235, plutonium-239, or uranium-233. Only one of these, uranium-235, occurs in nature and then only as about seven-tenths of one percent of natural uranium. By far the major fraction of cost and effort during the Manhattan Project was spent in producing concentrated uranium-235 and plutonium-239. Today, as then, there are two basic ways to produce concentrated fissile material. The first involves isotopic separation of uranium-235 from natural uranium; that is, an enrichment or increase is made in the concentration of uranium-235 compared to the nonfissile isotopes in natural uranium. The second method involves neutron bombardment (typically in a reactor) of the nonfissile isotopes uranium-238 and thorium-232, which occur in nature. This generates the fissile isotopes plutonium-239 and uranium-233 respectively. Since plutonium-239 is a different element from uranium-238, and uranium-233 is a different element from thorium-232, they can be separated by chemical methods that are simpler than those so far available for separating isotopes of the same element.

Today, a country wanting to acquire nuclear weapons need not start from scratch in order to produce the concentrated fissile material. Accepted civilian nuclear activities have, as a by-product, advanced many countries a long way towards the production of such material. Furthermore, since civilian nuclear activities overlap with military ones, they can provide a cover for further military advance.

When a government decides to develop nuclear weapons, a number of factors affecting the project—the extra costs, the extra time, the probability of success in that time interval, and the extra risks of exposure and counteraction—are largely determined by the "starting point," that is, the point after which an activity ceases to be "exclusively peaceful" within the meaning of various agreements on "atoms for peace." Even if a program to acquire nuclear weapons or the fissile material for such weapons violates no international agreements, there would be some point in keeping it inconspicuous or ambiguous, since regional adversaries or other governments might take counteraction. But a plain violation of an international agreement is a condition for the application of safeguards, and in any case would be riskier. What constitutes a violation depends, however, on the legal and other institutional conventions that define "exclusively peaceful" in the agreement.

The Acheson-Lilienthal proposal would have unambiguously outlawed uranium mines unless they were under international ownership. No one today suggests an international ban on all but internationally owned and controlled uranium mines. Some changes, however, are clearly called for in the present international conventions to assure that governments cannot legitimately come so close to obtaining nuclear weapons that "safe-guards against proliferation" would lose all meaning.

We will outline various pathways to a stock of concentrated fissile material and the various sets of legal restrictions that define the starting point of each pathway—that is, the beginning of the illegitimate part of the road to explosive nuclear material. Assuming a desired production rate of five to ten weapons a year, we estimate for each pathway the magnitude of the time and cost required to go the rest of the way to the needed fissile material. (We could make the numbers more precise only by specifying particular countries and circumstances.) In section 1.B we present the pathways in roughly increasing order of the cost, time, and difficulty of traversing them, their likelihood of success in a given time interval, and the risks of exposure and counteraction. A more detailed description of each of the elements in the pathways (section 1.C) follows the outline of the paths themselves. Section 2 provides a set of tables on world nuclear facilities and the sensitive nuclear materials they use or produce.

We omit pathways involving the outright purchase, gift, or theft of concentrated fissile material. More importantly, we do not treat the design and manufacture of the nonnuclear components and assemblies required to compress the fissile material to a supercritical mass. This is not because we think the design and production of implosion systems is a trivial task, as is sometimes suggested in discussions of individual or small groups of nuclear terrorists. On the contrary, we believe that the task is difficult; but given the information now public, and the improved materials, components, and equipment purchasable in the open market, it is certainly within the competence of a well-equipped and staffed government laboratory, and less difficult than acquiring the highly concentrated fissile material in facilities specifically designed to produce it. However, the essential point is that no agreement prohibits the use of high explosives to induce shock waves for the compression of metal; no such ban has ever been proposed nor would it be likely to be accepted or feasible to monitor. The monitoring of safe-guards in the programs of the IAEA and Euratom and of the various supplier nations has been directed at accounting for and controlling fissionable and other sensitive materials. The problem is that both the military and civilian uses of nonnuclear explosives are widespread in research and experimentation, in construction and in production. For example, chemical manufacturers have listed in their catalogue for many years explosive materials that are especially convenient and safe for inducing shock waves in metal as a regular industrial procedure for prehardening of metals, and for experiments in the rapid compression of heavy materials. Some nonweapon states have programs and sophisticated equipment for applying such procedures in the course of seismic research. We therefore focus on the acquisition of the concentrated fissile material. Here the possibilities for changing the conventions of control are more plausible.

B. Pathways to the Production of Weapon-Ready Nuclear Material

B.1. Pathway 0: Starting with weapon-ready nuclear material

This pathway starts legitimately with stocks of plutonium or highly enriched uranium in metallic form and with facilities for putting the metal into various shapes, or with the shaped metal itself. Stocking and shaping of concentrated fissile metal might be done for the accepted purpose of performing critical experiments or of fabricating highly concentrated fissile fuel. Governments in such a position essentially start with nuclear material ready to be inserted in a nonnuclear implosion system (weapon-ready nuclear material or WRNM; see section 1.C.8). The marginal or additional time and cost associated with this pathway, then, are nearly zero, as is shown in figure A-0. An example of a nonweapon state which may be at this starting point is Italy: at its Tapiro fast research reactor it has over 20 kg of a uranium-molybdenum alloy in which the uranium content is 98.5 percent, and that uranium itself is 93 percent uranium-235. Another example is Japan, which has performed spherical critical experiments with highly enriched uranium metal.

```
┌─────────┐
│         │
│  WRNM   │
│         │
└─────────┘
```

Figure A-0: Pathway 0

B.2. Pathway 1: Starting with materials in condition red

Here a nonweapon state could legitimately have an enrichment plant capable of quickly producing highly enriched uranium (see section 1.C.2) or a reprocessing plant capable of separating plutonium from spent fuel (see section 1.C.5); or it might receive from other governments the products of such plants in easily accessible form, such as highly enriched uranium hexafluoride or plutonium nitrate or plutonium dioxide. In chapter 1 of the text we have called this "condition red." The products of enrichment and plutonium separation plants we call "strategic special nuclear material" or SSNM (see section 1.C.8). Such material would, in general, have to be reworked in order to become weapon-ready (see figure A-1; for SSNM rework, see section 1.C.8). The facilities for such rework are in general quite simple. They involve no remote handling of radioactive material and today they are not illegitimate. From a starting point at which the SSNM is put into these facilities, the time required to produce weapon-ready nuclear material would be on the order of days to weeks. If such facilities were made illegal, the critical time would be extended by six months to a year, but since the plants are small and simple, they might be constructed in secret. The cost of traversing this path, if we include the cost of the facilities for reworking the SSNM, would be on the order of $1–10 million. Both Japan and West Germany are well started down this path. Japan has negotiated contracts with France and the United Kingdom to reprocess its spent oxide fuel and to return the recovered plutonium after a given time delay in a form to be specified by the Japanese. West Germany has received large shipments of highly enriched uranium hexafluoride which it uses for fabricating fuel.

Figure A-1: Pathway 1

B.3. Pathway 2: Starting with materials in condition orange

If it were illegal for a country to have an enrichment plant capable of quickly producing highly enriched uranium; or a plant for separating plutonium from hot spent fuel; or the products of such plants, it might still be legal for that country to receive from other countries or a multinational center fresh mixed plutonium and uranium oxide fuel (MOX) or highly enriched uranium (HEU) fuel for its light-water reactors (LWRs), its liquid-metal fast-breeder reactors (LMFBRs), or its high-temperature gas-cooled reactors (HTGRs). (The multinational reprocessing centers, current especially in proposals in 1975 and 1976, would have brought many countries to this particular starting point.) The country might instead be allowed to have a reprocessing plant that coprocesses spent fuel (that is, which removes the radioactive fission products leaving a mixture of uranium and plutonium [see section 1.C.5]). Japan has agreed with the United States to limit its Tokai Mura plant, at least for the first two years, to the coprocessing of spent fuel. This pathway, then, would begin with the fresh fuel or the coprocessed spent fuel. This starting point we have called "condition orange" in chapter 1 of the text. A fresh fuel reprocessing plant (see section 1.C.7) could be a quite small and inexpensive laboratory, since once again it involves no remote handling and the fissile material is quite concentrated in the fuel. The construction of the plant, which could be clandestine, might take several months to a year. However, the time elapsed from the seizure of the fresh fuel to the production of SSNM would be on the order of days to weeks. The SSNM would then have to be reworked to be ready for insertion in a weapon (rework facilities, discussed above under pathway 1, are again assumed to be illegal here), and this again might take days or weeks. The costs would be modest. For example, the cost of separating plutonium from the fresh MOX fuel would add only about a quarter of a million dollars to the $1–10 million needed for reworking the special nuclear material. This pathway is shown in Figure A-2.

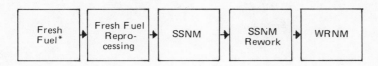

Figure A-2: Pathway 2

*That contains either plutonium or highly-enriched uranium.

B.4. Pathway 3: Starting with materials in condition yellow

If, in addition to the constraints that define the starting point of path-way 2, a nonweapon state agreed to forego the storage or use of fresh highly enriched uranium fuel or fresh plutonium fuel or coprocessed spent fuel, it could legitimately operate power reactors such as LWRs, AGRs, CANDUs, or Magnox, using slightly enriched or natural uranium fuel only once. It might also have research reactors that produce significant quantities of plutonium in the irradiated fuel. This pathway, then, would begin with the highly radioactive or hot spent fuel from power or research reactors. It corresponds to condition yellow in chapter 1 of the text.

A facility for extracting plutonium from the hot spent fuel, and in particular from the spent oxide fuel of pressurized light-water reactors (PWRs), which make up the vast majority of power reactors now being sold, is a substantially more difficult undertaking than the small laboratory that suffices for reprocessing the fresh MOX. For such a hot oxide reprocessing facility to have a comparable chance of being carried through successfully would require much more elaborate equipment, including facilities not only for remote operation but also for remote maintenance and repair in the event of malfunctions of the kind that have plagued the entire history of spent oxide reprocessing plants in the advanced industrial countries, especially in their initial operation. Its success, moreover, would depend strongly on the professional qualifications and experience of the people involved in the design, construction, and operation. In particular, it would depend very much on the existence of a core of technical personnel with actual professional experience in radioactive reprocessing. To be reliable, the plant would also require a substantial capacity, and we estimate that it is likely to take at least 18 to 24 months from the starting point of construction (which for this pathway would entail a violation of an agreement) before the plant could become operational. The enterprise would also be on a scale which would make clandestine operation much more difficult. (Recent proposals for a simple, quick version of such a plant have many flaws, including especially conceptual flaws.)

Apart from the time for constructing the plant, the elapsed time for the reliable production of material would be on the order of months to a year, measured from the seizure of spent fuel to the production of weapon-ready material. The costs would be at least $10 million but probably closer to $100 million. This pathway is shown in figure A-3.

Figure A-3: Pathway 3

*That contains either plutonium or highly-enriched uranium.

B.5. Pathway 4: Starting with construction of plutonium production and separation facilities "dedicated" to nuclear weapons

Aside from the weapon states, not many countries have large research reactors producing substantial amounts of plutonium, and in the future such reactors are not likely to be taken at face value as being solely for research. If alternatives for accomplishing the legitimate civilian ends of such research reactors were provided, and if a nonweapon state agreed to use such substitutes rather than to build a large plutonium production reactor, a government interested in weapons might elect to design and construct facilities specifically for the purpose of generating plutonium in reactors modeled on the early production reactors; and it might design and build a separation plant specifically to reprocess the spent fuel from such reactors. Such fuel would be of simpler geometry, would have more easily-processed cladding, and would be less intensely radioactive than power reactor fuel, especially the oxide fuel from light-water reactors. For such reasons the path to nuclear material through "dedicated" facilities has sometimes been thought of as cheaper and less time-consuming than the pathways we have described above. However, in this case all of the time, costs, and risks are the direct result of the decision to acquire nuclear explosive material—a very different case from the marginal or additional costs and risks that might be incurred if one had already in place nuclear facilities or materials for a nominal or actual civilian purpose. The costs of traveling pathway 4 might be at least the $10–100 million involved in pathway 3, and the time would be longer. Starting with the engineering and construction, it might take several years (Lamarsh[1] estimates 5 years) to complete the facilities and to produce the necessary material in a form ready for weapons. For many of the less industrial countries it would take considerably longer than that. In any case, the construction and operation of such facilities is unlikely to be kept secret from start to finish. Pathway 4 is illustrated in figure A-4. (We assume, as is explained in section 1.C.1, that a government embarking on pathway 4 or pathway 5 has natural uranium available.)

[1] Lamarsh, J. R., *On the Construction of Plutonium-Producing Reactors by Small and/or Developing Nations*, Report to the Library of Congress, Congressional Reference Service, 30 April, 1976, and *On the Extraction of Plutonium from Reactor Fuel by Small and/or Developing Nations*, Report to the Library of Congress, Congressional Reference Service, 19 July, 1976.

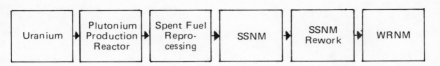

Figure A-4: Pathway 4

B.6. Pathway 5: Starting with construction of an enrichment plant "dedicated" to nuclear weapons.

Given the same legal constraints assumed in pathway 4, an alternative of the same sort would be to design and build a plant capable of producing highly enriched uranium. The feed for such a plant could be natural or slightly enriched uranium, both of which could be stocked legitimately. For the technologies presently available this would be a very difficult enterprise, taking at least five years, and very much longer for less industrial countries. It is unlikely that such a project could remain clandestine. The total cost, including that of all the constructed facilities, would be perhaps $100 million to $1 billion. As in the case of pathway 4, the political risks incurred by embarking on this pathway would be very much higher than in the earlier pathways. Pathway 5 is illustrated in figure A-5. It is conceivable that future technologies of enrichment might be cheaper, and if such advanced technologies were transferred without constraints, this would alter the time, costs, and risks.

Figure A-5: Pathway 5

*Capable of producing highly-enriched uranium.

C. *Specific Elements*

C.1. Uranium mining, refinement, and conversion

A major reason that nuclear energy is proposed as an alternative to fossil-fuel energy is the abundance of uranium ore throughout the world and the relatively low cost (per unit of useful energy) of mining and refining uranium. Table A-1 lists the results of a recent survey on world uranium reserves. "Reasonably Assured Resources" are mineral deposits where the extent of profitably recoverable ore is measured and samples have been analyzed. "Estimated Additional Resources" are potential sources that are likely to exist either as extensions of known fields or in

Appendix A

Table A-1

World Uranium Resources and Production--$50/LB U_3O_8

(Excludes Eastern Bloc Countries)

Thousand Tons U_3O_8

Country	Reasonably Assured	Estimated Additional	Production Planned 1978
North America	1,090	2,240	25.8
U.S.	840	1,370	19.3
Canada	237	853	6.5
Greenland (Denmark)	8	11	0
Mexico	6	3	0.02
Africa	740	260	12.4
South and SW Africa	452	94	8.8
Niger	208	69	2.4
Algeria	36	65	0
Gabon	26	13	1.2
C.A.E.	10	10	0
Somalia	8	4	0
Zaire	2	2	0
Madagascar	0	3	0
Australia	380	60	0.5
Europe	500	120	3.2
Sweden	391	4	0
France	67	57	2.85
Portugal	11	1	0.09
Spain	9	11	0.19
Yugoslavia	8	27	0
U.K.	0	10	0
Germany	3	5	0.1
Italy	2	1	0
Austria	2	0	0
Finland	4	0	0
Asia	60	31	0.2
India	39	31	0.2
Japan	10	0	0.03
Turkey	5	0	0
Korea	4	0	0
South America	80	20	0.3
Argentina	54	0	0.28
Brazil	24	11	0
Chile	0	7	0
Total (Rounded)	2,800	2,700	42.4

SOURCE: Uranium: Resources, Production and Demand, OECD/NEA and the IAEA, December 1977. Differences between this table and Table II of the text relate to the different price cutoff used and more up to date information.

regions where uranium is known to occur but has not yet been quantitatively assessed. These terms, used widely in the field to suggest the degree of probability that reserves will be found or made available at a given price, are not very rigorously defined. (See chapter 4.) They are fairly conservative definitions of actual resources. Additional resources of a more conjectural nature have been estimated for the U.S. and are in the process of being estimated for the rest of the world.

Currently worked and known deposits of uranium are concentrated in a few developed countries, such as the United States, Canada, Australia, and South Africa. Much of the less developed world, however, has been subject to only very partial or preliminary exploration compared to these countries. Thus there is a great potential for further uranium discoveries as sophisticated and intensive geological surveys are extended to new areas.

A few points should be noted about the economics of reserves. The accompanying table is for all uranium currently mineable for less than $50/lb. Clearly there is more or less uranium available at a higher or lower price. Since the cost of natural uranium is only a few percent of total nuclear power costs, the price of uranium could rise substantially with little effect on delivered electricity costs. There are countervailing pressures on uranium prices: on the one hand, known high-grade ore is being depleted; on the other hand, advancing technology is constantly adding more and more ore to the reserves (included at any given price cutoff), both by new mining and refining methods and by evolutionary improvements in old ones. This is especially important as there are several currently uneconomical ores of uranium which near-term efforts show promise of making economical. An example of this is the proposal to mine deeply buried but porous ores by pumping in a uranium solvent through a set of input wells and pumping out the solvent saturated with uranium from nearby output wells (a method similar to one already used to mine sulphur). A final key point on the economic aspect of reserve levels is that for a sufficiently high price, one can find uranium almost anywhere, even in seawater. Seeking nuclear independence, for example, India has mined ore in reserves which "did not exist" at the market price of uranium at that time. Thus for a small weapons-oriented program, the supply of uranium is unlikely to be a major constraint.

Currently mined reserves vary widely in nature and concentration. South African uranium is obtained from the tailings of gold mining operations, in which it has a concentration on the order of 0.02 percent. Sweden has huge amounts of low-grade bituminous shales (0.025 percent uranium). On the other side of the concentration spectrum, the United States, Canada, and Australia have large reserves in the few-tenths of a percent concentration range. Niger is estimated to have some 10,000 metric tons with a 2.8 percent concentration. A key point is that especially in some lower-grade ores, uranium is obtained as a by-product of gold, copper, or phosphate mining.

The technology involved in mining, concentrating, and refining uranium ores can be somewhat involved. It requires a large capital expense if the operation is to be conducted on an expanded scale with the intent of being economically competitive with world suppliers. However, the basic technology is well documented in the open literature.[2] Most countries, including developing ones such as Pakistan, have sufficient capability for small scale production of uranium suitable for use in nuclear reactors.

Moreover, the present trend is for international mining companies to participate with local governments in developing uranium resources. For example, Italy is planning to work with Bolivia in studying the commercialization of local uranium deposits. Companies from Japan, France, and West Germany have formed a Niger uranium exploration company.

Because there is a variety of sources and final states of uranium, no single precise method is suited to processing all ores. The basic steps, however, are always concentration, refining, and, if necessary, conversion into uranium metal or hexafluoride.

Concentration consists of separating uranium from the main bulk of the ore. The product is a crude uranium compound still contaminated by significant amounts (perhaps 1 to 5 percent) of inert material. Crude uranium concentrates may consist of oxides (U_3O_8), or salts such as sodium diuranate ($Na_2U_2O_7$), or ammonium diuranate [$(NH_4)_2U_2O_7$]. Concentration is usually carried out near where the raw ore is mined in order to reduce the bulk and weight of the ore that is shipped.

Refining consists of the removal of the remaining contaminants from crude uranium concentrates. This normally produces one of the oxides of uranium: UO_3, U_3O_8, or UO_2. The purification step removes from the uranium those trace elements which would interfere with its subsequent use in a reactor or isotope-separation plant. Once purified, the uranium oxide may either be used in a reactor directly, converted into uranium metal, or made into uranium hexafluoride feed for an enrichment plant.

C.2. Uranium enrichment

The purpose of enrichment is to obtain concentrated uranium-235. There are three major methods currently in use: gaseous diffusion, gaseous centrifugation, and the Becker or jet nozzle process. All exploit the difference in weight between uranium-235 and uranium-238 in uranium hexafluoride (UF_6), which is gaseous when heated. In each case, the effort involved in separating the isotopes is referred to as the separative work and is measured in Separative Work Units (SWU). Other enrichment technologies are under development, including enrichment through the use of lasers, and may come into use in the future. Only the technologies now in use are described below.

Gaseous diffusion: Separation by gaseous diffusion takes place because the lighter molecules in a gaseous mixture tend to move faster than the heavier molecules. Consequently they will strike container walls more often. If the walls have holes just large enough to allow passage of the individual molecules, without permitting bulk flow of the gas, more of the

2 Benedict, Manson, and Pigford, Thomas, *Nuclear Chemical Engineering*, (New York: McGraw-Hill, 1957).

lighter molecules than the heavy ones will diffuse through the wall. The theoretical separation that can be achieved in one application of this process is $\alpha = (M_2/M_1)^{1/2}$, where M_2 and M_1 are the molecular weights of the isotopic compounds. For UF_6 and the isotopes uranium-235 and uranium-238, α is only 1.00429. Consequently many stages are required to achieve a specified separation. This not only implies a slow process of enrichment, but also that there is a relatively large amount of uranium hexafluoride at various stages in the plant at any one time (all of which must be heated to remain gaseous).

Each stage in a diffusion plant (see figure A-6) requires a compressor to keep the feed material at a higher pressure than in the next stage. This causes a net flow in a single direction through the barrier. Large amounts of power are needed for these compressors and for the required heating elements.

Efficient operation of gaseous diffusion plants requires a relatively large facility, and construction of one is a major industrial and financial undertaking. The enormous power consumption of such a plant also creates a

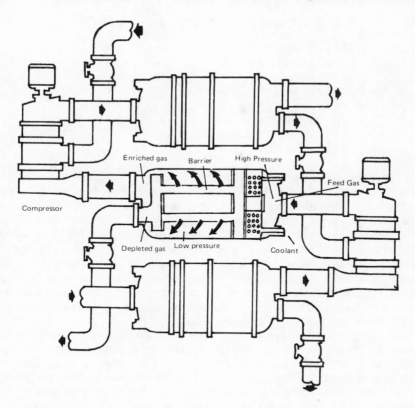

Figure A-6: Gaseous Diffusion Stage

problem, particularly for developing countries with a limited electrical network. The U.S. gaseous diffusion complex, for example, requires 6000 MW of electricity for its three facilities when operated at full capacity. This is approximately the power consumption of the entire state of Minnesota.

The technological difficulty in constructing a gaseous diffusion plant lies primarily in the construction of the barriers. Both barrier geometry and material are highly classified.

Gaseous centrifugation: The gas centrifuge employs a rotating cylinder which imparts a centrifugal force driving the UF_6 molecules toward the outer radius of the cylinder. However, the thermal velocities of the molecules tend to keep the gas molecules evenly distributed throughout the available volume. Since the latter effect is larger for the lighter uranium-235 molecules, they tend to concentrate towards the center. A given stage in centrifugation can have a relatively large separation factor but invariably has a smaller flow rate than a diffusion stage. For a centrifuge operating isothermally, the maximum separative capacity of the machine is

$$\text{Max } SW = \pi\rho D(Z/2) \; [M_2 - M_1) \; \mu^2/2RT]^2$$

where SW = separative work
ρ = gas density
D = diffusivity
$M_2 - M_1$ = isotopic mass difference
Z = bowl length
μ = peripheral velocity
T = absolute temperature
R = gas constant.

Note that the separative work is here a function of the difference in molecular weights and not their ratio as with the diffusion process. This separative work equation also indicates the importance of a rapidly spinning cylinder (4th power of the peripheral velocity) and a low temperature.

In 1960 results were published[3] on centrifuges approximately one foot long, spinning at 350 m/sec at a temperature of 33°C. These parameters yielded separation factors of 1.1–1.2 (compared to the maximum for diffusion of 1.004). Thus only relatively few centrifuges need to be connected in series to achieve substantial changes in concentration. However, because of low flow rates, large numbers of centrifuges in parallel are required for a practical level of throughput.

The technical difficulties of exploiting the centrifuge process lie in the design and production of lightweight centrifuges that have a low friction coefficient (and low power consumption). The lightweight bowls must be able to withstand the centrifugal forces derived from the high peripheral speeds. Several of the highly industrialized countries have produced successful machines with component reliability of 98 to 99 percent. In the free

[3] Zippe, G., "The Development of Short Bowl Ultracentrifuges," Research Laboratories for the Engineering Sciences, University of Virginia, 15 June 1960.

world, current involvement in centrifuge technology is primarily in the U.S., in the trilateral agreement countries of Urenco/Centec (UK, West Germany, and the Netherlands), and in the Power Reactor and Nuclear Fuel Development Corporation (PNC) of Japan.

The Becker nozzle process: Figure A-7 presents a schematic of the Becker nozzle process being developed in Karlsruhe, West Germany (and possibly in South Africa), which has progressed to the stage where pilot production plants can be built. Claims have been made that further improvements being investigated may produce a reduction in the costs of this system.[4]

Uranium hexafluoride gas (mixed with hydrogen or helium) is pumped through a long slit, forming a rapidly moving sheet of gas. The gas strikes a curved wall, bending the sheet through 180 degrees. Centrifugal forces

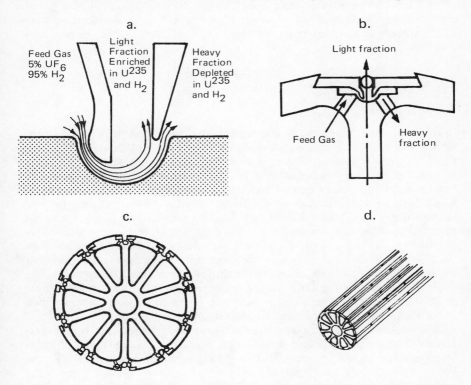

Schematic (a & b) shows flow of gas through a separation element in the Becker nozzle process. Multiple elements are arranged along a cylinder (c & d) to form one of many stages in the process train.

Figure A-7: Nozzle Schematic

[4] Smith, David, "What Price Commercial Enrichment?" *Nuclear Engineering International* 19, no. 218 (July 1974): 582.

then carry the heavier uranium-238 (which makes up 99.3 percent of the gas) to the outer surface of the sheet, where a knife-like barrier pares it off. The lighter fraction, now slightly enriched in uranium-235, is routed through hundreds of additional stages to reach the desired level of enrichment. Each stage has an α of about 1.01–1.02, intermediate between those of diffusion and centrifugation.

Advantages of the Becker nozzle process are that it does not require the hard-to-make porous barriers of a gaseous diffusion plant and that it avoids the highly stressed moving parts of an ultracentrifuge. One disadvantage of the nozzle process is that it consumes more electric power than gaseous diffusion and at least 10 times more power than an equivalent centrifugation plant. This disadvantage wanes, however, where electric power is cheap (even if inconveniently located), as in Brazil.

In South Africa a pilot enrichment plant which is thought to be based on the nozzle technique is operating. Plans have been announced to construct a large-scale plant (perhaps 5 million SWU/year) for operation in 1987.

C.3 Nuclear reactors

Power reactors

Nuclear power reactors use the heat created by fission to produce steam which is then used to create electricity much as in any other thermal power plant (driven by coal or oil for example). The principal parts of a reactor are its fuel, coolant, and moderator, each of which to varying degrees differs between one type of reactor and another. The fuel is generally uranium, which in some types of reactors has been "enriched" by increasing the percentage of fissile uranium-235. It can also be enriched by adding plutonium or uranium-233. The moderator is a material, either graphite or water, designed to slow neutrons produced by fission down to "thermal" rather than "fast" velocities. This increases the neutrons' efficiency for subsequent fissions and reduces loss of neutrons through uranium-238 capture. The coolant is a circulating fluid, either a gas or liquid, which transfers heat from the reactor core to a water boiler in order to produce steam which in turn is used in a conventional fashion to create electricity.

The most important power reactors are the graphite-moderated gas-cooled reactor (Magnox and graphite-gaz reactors); the two types of light-water reactor (LWR), pressurized- and boiling-water reactors (PWR and BWR); the CANDU pressurized heavy-water reactor (PHWR); the high-temperature gas-cooled reactor (HTGR); and the liquid-metal fast-breeder reactor (LMFBR). All of these reactors, in addition to producing power, either produce plutonium or use highly enriched uranium or do both—in any case presenting a proliferation potential. The characteristics of each of these major reactor types are summarized in table A-2.

Magnox and graphite-gaz reactors: The Magnox reactor uses natural uranium metal fuel rods clad in "Magnox," a magnesium alloy. These are inserted in graphite blocks and cooled by compressed carbon dioxide. The French call their variant "graphite-gaz" for natural uranium graphite-

Table A-2

Reactor Characteristics

Reactor Type	Usual Fuel	Coolant	Moderator
Magnox and Graphite-gaz	natural uranium metal	compressed carbon dioxide	graphite
LWR (BWR and PWR)	slightly-enriched uranium dioxide	light water	light water
CANDU PHWR	natural uranium dioxide	heavy water	heavy water
HTGR	highly-enriched uranium	helium	graphite
LMFBR	plutonium dioxide	liquid sodium	none
Plutonium Production Reactor	natural uranium metal	air	graphite

moderated and gas-cooled. The British and French have produced many such reactors for their domestic markets and have exported three. The British AGR (advanced gas-cooled reactor) is an improved version using slightly enriched uranium in oxide form. Enrichment allows longer burnup periods between fuel changes, and oxide fuel can stand higher operating temperatures, which enable higher thermal efficiency.

Light water reactors: Most power reactors are light-water reactors (light water is normal water, it is called light water to contrast it with heavy water), of which there are two types, pressurized- and boiling-water reactors. In both cases, light water is used as the moderator and coolant and along with the fuel assembly is contained in a pressure vessel containing the entire reactor core. In a pressurized-water reactor the coolant water remains liquid due to the high pressure maintained and creates steam by in turn heating less highly pressurized water in a separate boiler. In a boiling-water reactor, lower internal pressure is maintained, and the coolant water turns into steam itself and drives the generator directly. In either case, due to the degree of neutron absorption by light water, enriched uranium must be used for efficient operation.

CANDU pressurized heavy water reactor. The CANDU PHWR differs from the PWR in several ways. First, its coolant and moderator is heavy water—made up with the relatively rare heavy isotope of hydrogen, deuterium. The chief significance of this is twofold: since heavy water absorbs many fewer neutrons than light water, a CANDU reactor will work efficiently with natural uranium; on the other hand, heavy water itself requires sophisticated facilities to be produced in quantity, although not as sophisticated as those required to produce enriched uranium. A second major difference is that rather than containing the water and fuel assemblies in a single pressure vessel, a CANDU individually surrounds each of hundreds of "fuel tubes" with a heavy-water coolant "pressure tube." As in the PWR, the CANDU coolant is kept pressurized as a liquid and in turn heats water in a separate boiler. A separate system of heavy water—surrounding but insulated from the pressure tubes—acts as the moderator and is kept at room temperature and approximately normal pressure. CANDUs and LWRs differ also in their refueling. The CANDU (and the Magnox and the graphite-gaz), which is not contained in a pressure vessel, can be continuously fueled while operating. LWRs must be shut down for refueling.

High-temperature gas-cooled reactor: While the moderator and coolant of the HTGR are much the same as with the Magnox and the graphite-gaz, the fuel and operation are quite different. Higher operating temperatures and the use of helium rather than carbon dioxide as the coolant gas increase thermal efficiency. The higher temperatures are possible since the highly enriched uranium fuel is in the form of oxide or carbide pellets surrounded by layers of refractory carbon and silicon carbide. Due to the high enrichment of uranium, substantial amounts of thorium-232 pellets may be mixed in which absorb neutrons and turn into fissile uranium-233. This decreases somewhat the uranium requirements of the reactor.

Liquid-metal fast-breeder reactor (LMFBR): The LMFBR is generally fueled by plutonium. It has no moderator to slow neutrons down—hence the "fast" part of its name. Plutonium fissions caused by fast neutrons release a relatively large number of neutrons. By surrounding the plutonium with blankets of "fertile" uranium-238, neutrons not used in the chain reaction are absorbed, "breeding" more fissile plutonium (plutonium-239) than is being used up. Reprocessing and recycling of the bred plutonium is a necessary part of the LMFBR fuel cycle. Due to the need not to moderate the neutrons in the reactor and due to the high temperatures involved, a liquid metal (generally sodium) is used as the coolant. Since the heated sodium is used to create steam, there must at some point be a heat exchanger which contains both liquid sodium and water. Since liquid sodium tends to react violently with water even the slightest leaks in the heat exchangers can cause serious problems. (Many LMFBRs have had trouble operating due to heat exchanger leaks.)

Plutonium-production reactors

The plutonium-production reactor is optimized for producing fissile plutonium-239 and generally will not produce any power. The characteristics of this reactor are summarized in table A-2. An example of a plutonium-production reactor is the French G1. The G1 is essentially a simplified version of the graphite-gaz design optimized for plutonium production. Like the graphite-gaz, it is gas-cooled (air rather than compressed carbon dioxide), graphite-moderated, and fueled by natural uranium metal. Some plutonium-production reactors use heavy water as the moderator and coolant. The U.S. production reactors at Savannah River are an example of this type.

Research reactors

Research reactors are used mainly as a source of neutrons. They come in such a wide variety, using different moderators, coolants, and uranium enrichments, that it will not be possible to describe a typical research reactor. Most of them do not use or produce fissile material in quantities that are significant from the standpoint of a weapons program, but those that do should cause us some concern. We shall focus therefore only on those that make substantial amounts of fissile material accessible either in their fresh or spent fuel.

Research reactors that have high power levels and use natural or slightly enriched uranium for fuel produce significant quantities of separable plutonium in the spent fuel. Table A-3[5] lists the countries that have such research reactors. They produce one to 12 kg of plutonium per year, except for Canada, whose three research reactors together produce 77 kg per year, or enough plutonium for about 15 bombs. Most of the countries listed have power reactors, any one of which could generate more plutonium than any of their research reactors. India's and Israel's research reactors, however, are of particular concern, since they are subject to no safeguards. Research reactors of this sort are difficult to distinguish from reactors "dedicated" to the production of plutonium and therefore can provide a civilian front for the production of plutonium. (The first French plutonium-production reactor, the G1, had essentially the same design as the Brookhaven Graphite Research Reactor.)

Research reactors that use as fuel substantial amounts of highly enriched uranium (HEU)—that is, uranium with more than 20 percent uranium-235—are a problem because this fresh fuel could be used quite directly in order to make nuclear weapons. Table A-4 lists research reactors that use HEU. What one would like to have is precise information on the stocks of unirradiated, highly enriched uranium in each country. Unfortunately such data are not available. However, one can use table A-4 to get some idea of the sizes of these stocks. For reactors with low power outputs (less than about 10 kW) one can assume that the HEU remains essentially unirradiated even after being in the reactor for a long time. Reactors with high power outputs (greater than about 1 MW) will burn up fuel so quickly that replacement fuel assemblies will have to be kept on hand at all times. As a rough approximation one can assume that this amount of material will be between 1 and 2 full core loadings.

Research reactors that use fuel with an enrichment between 10 and 20 percent present a relatively small proliferation hazard since they neither produce large amounts of plutonium nor have fuel that is directly usable in a nuclear weapon. Since the function of research reactors is basically to provide neutrons, reactors that use 10 to 20 percent enriched fuel may provide a proliferation-resistant substitute for most of the reactors listed in tables A-3 and A-4.

---

[5] Both tables A-3 and A-4 are drawn from table A-9. Table A-3 is believed to be complete. No technical information, however, is known for some of the reactors listed in table A-9. A number of the reactors for which technical information is known may have been converted to highly enriched uranium. As a result, table A-4 may well be incomplete.

Table A-3

Research Reactors which Produce Substantial Amounts of Plutonium

Country	Reactor Name	Uranium Enrichment	Power Output	Moderator	Start Year	Estimated kg Production per Year*	At the Beginning of:				
							1970	1975	1980	1985	1990
Belgium	BR-1	nat.	4 MW	Graphite	1956	1	14	19	24	29	34
	BR-3/VN	7%	40.9 MW	20%-100% D_2O**	1969	≈1	1	6	11	16	21
Canada	NRX	nat.	40 MW	D_2O	1947	12	276	336	396	456	516
	NRU	nat.	200 MW	D_2O	1957	60	780	1080	1380	1680	1980
	WR-1	2.4%	60 MW	D_2O	1965	≈5	25	50	75	100	125
FRG	FR-2	nat.***	12 MW	D_2O	1961	4	40	65	90	115	140
India	CIRUS	nat.	40 MW	D_2O	1960	12	120	180	240	300	360
Israel	IRR-2	nat.	26 MW	D_2O	1964	8	48	88	128	168	208
Italy	ESSOR	nat.****	40 MW	D_2O	1967	≈3	9	24	39	54	69
Japan	JRR-3	nat.	10 MW	D_2O	1962	3	24	39	54	69	84
Norway	HBWR	1.5%	20 MW	D_2O	1959	3	33	48	63	78	93
Switzerland	DIORIT	nat.	30 MW	D_2O	1960	9	90	135	180	225	270
Taiwan	TRR	nat.	40 MW	D_2O	1973	12	--	24	84	144	204
Yugoslavia	R-A	2%	10 MW	D_2O	1959	≈1	11	16	21	26	31

* Assumes 1 kg of plutonium produced for every 1,000 MWD of irradiation in a natural uranium reactor. The reactor is assumed to operate 300 days per year. The plutonium production is inversely proportional to the fuel enrichment. The percentage of the plutonium that is fissile depends on the fuel burnup. In most cases, the percentage of fissile plutonium is between 85% and 95%.

** D_2O stands for deuterium oxide, which is heavy water.

*** Apparently modified to 1.5%-2% enriched uranium in 1966. Power output increased to 44 MW and yearly plutonium production increased to about 5 kg.

****Core consists of 7 kg of 90% enriched uranium and 670 kg of natural uranium.

Table A–4

Research Reactors with Fuel Enrichment Greater than 20%

Country	Reactor Name	Amount of Uranium in Core	Power Output	Uranium Enrichment
Argentina	RA-3	4.12 kg	5 MW	90%
Australia	HIFAR	3.09 kg	10 MW	93%
	Moata	3.32 kg	10 kW	90%
Austria	Astra	4.46 kg	5 MW	90%
Belgium	BR-02	1.63 kg	500 W	90%
	BR-2	5.56 kg	100 MW	90%
Canada	MNR	4.33 kg	2 MW	90%
	NRX	4.92 kg	40 MW	93%*
	PTR	2.90 kg	100 W	90%
Colombia	IAN-R1	2.42 kg	20 kW	90%
Denmark	DR-2	5.00 kg	5 MW	90%
	DR-3	3.09 kg	10 MW	93%
FRG	Adibka-1	1.24 kg	10 W	93%
	FMRB	3.58 kg	1 MW	90%
	FRG-2	7.00 kg	15 MW	90%
	FRJ-1 (Merlin)	3.98 kg	10 MW	80-90%
	FRJ-2 (Dido)	3.09 kg	23 MW	80%
	Sneak	variable	1 kW	20, 35, & 93%
India	Apsara	7.61 kg	1 MW	46%
Israel	IRR-1	4.78 kg	5 MW	90%
Italy	Avogadro RS-1	4.78 kg	7 MW	90%
	Essor	7.00 kg	40 MW	90%
	Galileo Galilei	5.56 kg	5 MW	89.9%
	RITMO (RC-4)	8.22 kg	100 W	90%
	ROSPO	18.92 kg	Negl.	89.9%
	Tapiro	23.14 kg	5 kW	93.5%
Japan	JMTR	6.83 kg	50 MW	90%
	JMTRC	6.83 kg	10 W	90%
	JRR-2	4.22 kg	10 MW	90%
	JRR-4	3.33 kg	1 MW	90%
	KUR	3.77 kg	5 MW	90%
	UTR-10 KINKI	3.37 kg	0.1 W	90%
Netherlands	Athene**	2.01 kg	10 kW	93.3%
	HFR	4.67 kg	30 MW	90%
	HOR	3.89 kg	2 MW	90%
	LFR	4.85 kg	10 kW	90%
Pakistan	PARR	4.78 kg	5 MW	90%
Poland	Anna	28.57 kg	100 W	21%
	Maryla	6.94 kg	10 kW	10-36%
Rumania	Triga	55 kg***	5 MW	93%
South Africa	Safari-1	3.73 kg	20 MW	90%
Spain	Coral-1	22.01 kg	10 W	90%
Sweden	R-2	6.89 kg	50 MW	90%
	R2-0	4.00 kg	1 MW	90%
Thailand	TRR-1	3.33 kg	1 MW	90%
Turkey	TR-1	4.78 kg	1 MW	90%

*This is a natural uranium reactor, but it is using the highly enriched uranium in order to compensate for reactivity loss due to experiments. This reactor also uses 3.56 kg of plutonium for this purpose.
**Shutdown 1973.
***This is the total amount being exported to this reactor. It is not known how much material is required for a core loading.

C.4. Nuclear reactor fuel fabrication

Once uranium has been mined, purified, and enriched to the desired level, the next step is fuel fabrication. In general there are several activities required to prepare the enriched uranium or the plutonium for insertion into a reactor core: processing of fuel materials, cladding of processed fuel, and preparation of fuel assemblies.

Although it is possible to use uranium in metal form (and there are a number of such reactors operational today), it turns out to be much more desirable to use the ceramic uranium dioxide (UO_2). This form exhibits better high-temperature stability, radiation resistance, and is chemically inert to hot water. When formed into pellets by the conventional ceramic-working procedures (extrusion, cold pressing, etc.) it also tends to contain a large fraction of the fission-product gasses and to minimize volume. Unfortunately the other desired fuel characteristic, thermal conductivity, is a somewhat weak point for UO_2.

Most fuels must be assembled into rods and clad with an appropriate material to preserve a geometric structure in the reactor core and to prevent interaction with the coolant. In light-water or heavy-water reactors with UO_2 fuel, the common cladding is stainless steel or zircaloy. The latter has advantages in lower neutron-capture cross section, but it is more expensive and less resistant to thermal stress.

When the coolant is sodium, as in the breeder, stainless steel is required. Gas-cooled reactors that do not operate at high temperatures and that use metallic fuel may use aluminum or magnesium cladding (such as the British Magnox). A special case is the high-temperature gas-cooled reactor (HTGR) fueled by highly enriched uranium carbide pellets. The uranium carbide must be enclosed in a graphite shell as are the thorium pellets used as fertile material. These are then combined in a graphite matrix.

Once the fuel rods have been prepared they may be combined into fuel assemblies which satisfy the desired geometry for a particular reactor. Unless plutonium fuel is involved, mechanical assembly is relatively straightforward. The radioactivity of plutonium-dioxide fuel necessitates use of glove boxes, shielding, and/or automated or remotely operated process equipment. No commercial-scale fabrication facilities for plutonium fuel have been constructed; thus future costs are highly uncertain. Quotes by existing fabricators have been rising rapidly, multiplying by a factor of over 3 between the beginning of 1971 through the end of 1974.[6] In its major analysis of the value of implementing a plutonium fuel cycle, the Edison Electric Institute estimated that the cost of fabricating plutonium fuel would be $302 per kilogram, 4.5 times its estimate of the cost of fabricating uranium-oxide fuel.[7]

---

[6] Stoller, F., et al., "Report on Reprocessing and Recycle of Plutonium and Uranium— Task VII of the EEI Nuclear Fuels Supply Study Program," Edison Electric Institute, March 1976.

[7] See table A-6.

C.5. Reprocessing

The term "reprocessing" is used to refer to the complex series of mechanical and chemical steps used to separate the uranium and plutonium contained in spent reactor fuel from the fission products and fuel cladding. Reprocessing may be considered useful by those who would like to (a) recover the remaining uranium-235; (b) recover the plutonium for reuse in thermal reactors (either light-water or heavy-water reactors); (c) recover the plutonium for use in a breeder reactor; or d) obtain plutonium-239 or uranium-233 for weapons purposes. Plutonium-239 can be substituted for uranium-235 in reactor fuel, which provides the basis for arguing in favor of recovery of plutonium from spent fuel.[8] On the other hand, reprocessing would have environmental, technical, and political risks that militate against its use in civilian nuclear programs. The possible economic benefits and associated risks (including risks of technical failure) are discussed in the text of this book. The following presents a brief technical description of the most commonly used reprocessing technology, and discusses the arguments for and against reprocessing.

Irradiated fuel removed from presently operating power reactors contains varying amounts of the useful fissile or fertile isotopes of uranium and plutonium as the fission by-products. If a thorium-based fuel cycle were to be developed in the future, spent fuel containing thorium and uranium-233 would also be produced.

During the operation of present reactors, the sequence of reactions illustrated in figure A-8 takes place. A comparable diagram for fuel using thorium as the fertile material is also provided.

The need for fuel replacement is dominated by fission-product poisons (isotopes that absorb the neutrons needed for the chain reaction) rather than by exhaustion of readily fissionable isotopes. The spent fuel from light-water reactors typically contains about 0.7–1.0 percent of fissile plutonium and somewhat less than 1.0 percent of uranium-235. Spent fuel from CANDU reactors contains much lower fractions of fissile isotopes, typically about 0.4 percent plutonium and 0.2 percent uranium.

A 1000 MWe LWR produces about 30 MTU (metric tons uranium) of spent fuel a year, containing about 300 kg of plutonium and about 900 kg of fission products.

Given this general background on the physics involved, the discussion can turn to a description of the reprocessing activities. Broadly speaking, there are five main steps involved in fuel reprocessing: cooling; head-end treatment; separation/extraction; conversion of plutonium nitrate to pluto-

---

8 A number of countries have built or have expressed an interest in obtaining reprocessing facilities, or in having their spent fuel reprocessed on contract in other nations, ostensibly because of the fuel value of plutonium. These contracts, in general, call for the return of the separated plutonium to the originating country. In addition to the known facilities listed in table A-12, Taiwan and South Korea have shown interest in having a reprocessing plant.

LEGEND:

──────▶ Major reaction path

─ ─ ─ ─▶ Minor reaction path

$\xrightarrow{(n,\gamma)}$ Neutron capture with gamma ray emission

$\overset{\longleftarrow}{(n,2n)}$ Neutron capture resulting in the emission of 2 neutrons

$\beta\uparrow$ Radioactive decay resulting in beta particle emission

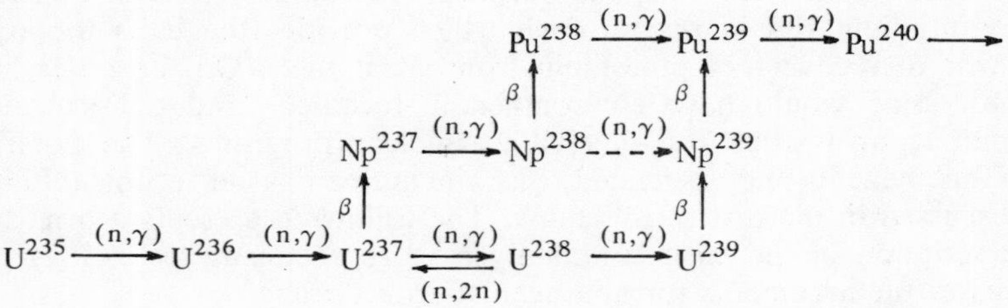

$$Pu^{238} \xrightarrow{(n,\gamma)} Pu^{239} \xrightarrow{(n,\gamma)} Pu^{240} \longrightarrow$$

$$\beta\uparrow \qquad \beta\uparrow$$

$$Np^{237} \xrightarrow{(n,\gamma)} Np^{238} \dashrightarrow Np^{239}$$

$$\beta\uparrow \qquad \beta\uparrow$$

$$U^{235} \xrightarrow{(n,\gamma)} U^{236} \xrightarrow{(n,\gamma)} U^{237} \underset{(n,2n)}{\overset{(n,\gamma)}{\rightleftarrows}} U^{238} \xrightarrow{(n,\gamma)} U^{239}$$

Actinide Buildup in Uranium

$$Cm^{242} \xrightarrow{(n,\gamma)} Cm^{243} \xrightarrow{(n,\gamma)} Cm^{244}$$

$$\beta\uparrow \qquad \beta\uparrow$$

$$Am^{241} \xrightarrow{(n,\gamma)} Am^{242} \dashrightarrow Am^{243} \xrightarrow{(n,\gamma)} Am^{244}$$

$$\beta\uparrow \qquad \beta\uparrow$$

$$Pu^{239} \xrightarrow{(n,\gamma)} Pu^{240} \xrightarrow{(n,\gamma)} Pu^{241} \xrightarrow{(n,\gamma)} Pu^{242} \xrightarrow{(n,\gamma)} Pu^{243}$$

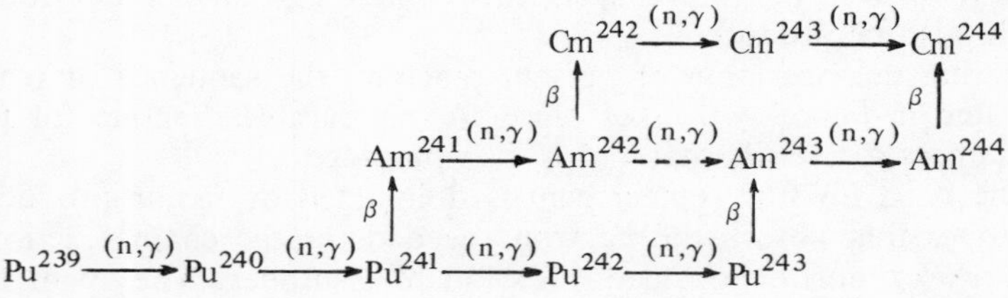

Actinide Buildup in Plutonium

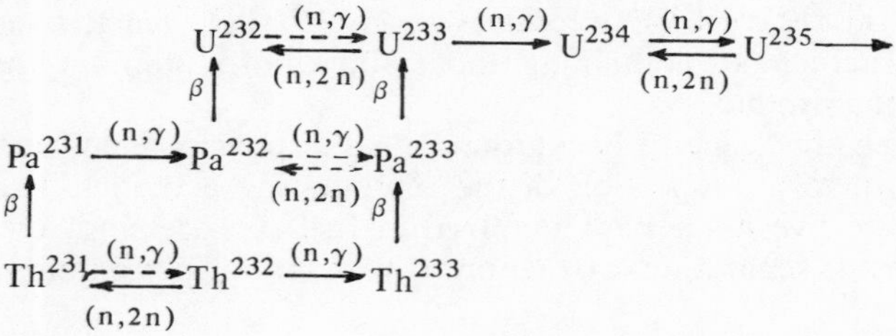

$$U^{232} \underset{(n,2n)}{\overset{(n,\gamma)}{\dashleftrightarrow}} U^{233} \xrightarrow{(n,\gamma)} U^{234} \underset{(n,2n)}{\overset{(n,\gamma)}{\rightleftarrows}} U^{235} \longrightarrow$$

$$\beta\uparrow \qquad \beta\uparrow$$

$$Pa^{231} \xrightarrow{(n,\gamma)} Pa^{232} \underset{(n,2n)}{\overset{(n,\gamma)}{\dashleftrightarrow}} Pa^{233}$$

$$\beta\uparrow \qquad \beta\uparrow$$

$$Th^{231} \underset{(n,2n)}{\overset{(n,\gamma)}{\dashleftrightarrow}} Th^{232} \xrightarrow{(n,\gamma)} Th^{233}$$

Actinide Buildup in Thorium

Figure A-8: Actinide Buildup in Uranium, Plutonium, and Thorium nium oxide; and storage of radioactive wastes and conversion of these to forms suitable for long-term storage.

Cooling: The cooling phase involves reduction in the temperature of the spent fuel and, more importantly, a decrease in radioactivity. This lessens the difficulty of the following reprocessing steps by: reducing the radiolytic damage to the reprocessing fluids; reducing the number of fission products (those with short half-lives decay) or impurities that have to be removed; and allowing radioactive decay of certain heavy elements that are chemically inseparable to concentrations where they no longer present a problem.

Typically 100 days of cooling will be a minimum for any fuel which is to undergo separation by solutions. After 100 days the overall beta and gamma activity will be one ten-thousandth of the level in the first hour after removal from the reactor. Only about a dozen isotopes contribute significantly to the radioactivity at that point. For example, uranium-237, which is highly radioactive, will have decayed to neptunium-237, which can be removed by relatively straightforward chemical means. Furthermore, neptunium-239 will have almost completely decayed into fissile plutonium-239, thus increasing the amount of the plutonium recovered. In actual practice, cooling times will be much longer than 100 days. Past difficulties with attempts to reprocess highly radioactive uranium oxide fuel have led planners to base designs for new plants on one to two years of cooling.

It should be noted that thorium fuels complicate the reprocessing task somewhat. Although the half-lives of protactinium-233 (which becomes fissile uranium-233) and thorium-234 are acceptably short, a delay to permit the protactinium-233 to decay would allow thorium-228 with a 2-year half-life to build up. It may be necessary to separate the protactinium-233 from the thorium early in the cooling period. This will be likely to complicate handling.

Head-end processes: Once cooling has been accomplished, the next step is to remove the spent fuel from its cladding, dissolve it, and clarify the solution of solids that would interfere with the solvent extraction process that follows. However, there are a wide variety of cladding materials, depending on the type of fuel and the coolant used in the reactor. Thus a single reprocessing facility may not be able to handle all types of fuels without extensive modification. Table A-5 gives some typical cladding and fuel combinations.

The fuel rods must be handled remotely inside concrete "caves" which are several meters thick and have special radiation-absorbing windows. The remote manipulators can then be used to perform one or more of the decladding processes. Uranium oxide fuel rods are mechanically chopped into sections, then dropped into acid which dissolves the fuel but not the cladding "hulls." At present, the hulls are placed in temporary storage at all reprocessing plants. Eventually they will need to be packaged and transferred to final storage. The solution is then clarified by centrifuges and passed to the next step. During the decladding and dissolution steps, certain of the fission products enter the environment. In particular, all proposed plants plan to allow krypton-85 and tritium to escape into the atmosphere without entrapment.

Separation/extraction: Once the fuel has been dissolved and clarified it can then undergo the third major step: extraction and separation of the uranium and plutonium from the fission by-products. The only separation method in wide commercial use today is the Purex process. There are basically three steps in the process: extraction of the uranium and plutonium from the fission products; partitioning the uranium from the plutonium; and stripping either the uranium or plutonium from its accompanying organic solvent. These steps may need to be repeated several times until the desired degree of purity and reduction in radioactivity is obtained. Typically the decontaminated product will have a factor of 10^7 to 10^8 fewer fission products than the initial feed.

Table A-5

Fuel Cladding

Fuel	Coolant/Reactor	Acceptable Cladding
Uranium dioxide	PWR or BWR	Stainless Steel or Zircaloy
Uranium dioxide	Sodium	Stainless Steel
Metallic uranium	Gas (not high temperature)	Aluminum or magnesium
HEU	HTGR	Silicon carbide coated pellets, graphite block matrix, stainless steel clad

Figure A-9 illustrates the Purex process. This chemical approach has evolved somewhat over the last three decades and now uses improved solvents (such as tributyl phosphate [TBP]). However, it is still basically a complex, expensive process relying on large facilities for its tall columns and multiple stages.

The fundamental principle of the Purex process is that uranium and plutonium have higher oxidation (valence) states than the fission products. This allows them to be more readily dissolved in a variety of organic solvents than the fission products. A tall column (20 stories at the British Windscale) is used with a heavy aqueous solution such as nitric acid introduced at the top, the clarified feed of dissolved spent fuel in the middle, and a lighter organic solvent at the bottom. Gravity alone, an impulse piston, or a series of mixer-settler banks may then be used to bring the liquids into contact. (These devices are called contactors.)

It is desirable to avoid pumps or other such devices because of the difficulty in performing repairs and maintenance on radioactive equipment. Furthermore, all these processes must be carried out by remote operation in heavily shielded "canyons" of concrete. Reprocessing plants built for the American and British nuclear-weapons programs were designed so that the apparatus in these shielded areas would never need maintenance or repair. Reliability was dependent on extraordinary design measures, very high quality materials, and in some cases on the construction of two complete primary separation plants.

Once the first extraction step is complete, the organic solution of U^{+6} and Pu^{+4} is passed to a partitioning stage in which fresh organic solvent and nitric acid with a reducing agent are introduced into another column. Gravity or the other means mentioned above is used to combine the immiscible liquids. As the dense nitric acid passes down through the organic solution, the reducing agent converts the Pu^{+4} to Pu^{+3} which is not soluble in the organic liquid. The uranium moves upward in the organic solvent and passes to a uranium stripping step while the plutonium leaves from the bottom for an additional extraction and stripping sequence.

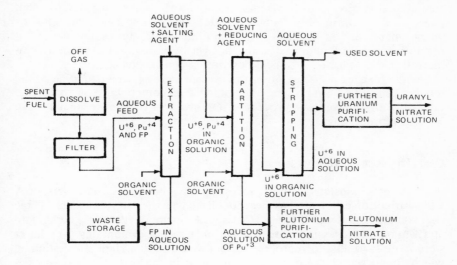

Figure A-9: Purex Process Flow Sheet

An alternative to the partitioning step is coprocessing. After co-decontamination in the extraction step, the plutonium and uranium can be further purified together in the organic solution. Once sufficient fission products are eliminated, the organic solvent can be stripped and the decontaminated uranium and plutonium nitrate will be left. It would be necessary to increase the ratio of plutonium to uranium from the 1 percent in the output of coprocessing to the 3 to 4 percent required for typical light-water fuel or the 10 to 30 percent required for typical fast-reactor fuels. It is technically unclear how this might be done without creating the potential for rapid production of weapon-grade plutonium.

The next step in the basic Purex process is to strip the uranium (or plutonium) out of the organic solvent. This is performed in a third column in which nitric acid flows down over the TBP but without the addition of the salting agents that were in the extraction and partitioning columns. The result is that the uranium is back-extracted as uranyl nitrate in aqueous solution. The solvent can be recovered and purified for reuse.

Conversion to plutonium oxide: The plutonium nitrate is then converted to plutonium oxide. Oxalic acid is added to the plutonium nitrate solution, forming plutonium oxalate which precipitates and can be filtered from the solution. The plutonium oxalate is then roasted at high temperature and it decomposes to form plutonium oxide. This conversion makes the plutonium safer to transport since it is in a solid instead of a liquid form. It also puts the plutonium in the chemical form in which it will be reused.

Conversion and storage of radioactive wastes: Reprocessing produces large volumes of liquid and solid wastes of varying degrees of radioactive contamination. Programs for managing these wastes are in the development phase in all countries. No country has decided on a specific plan for their management, and a crucial element in all plans, the solidification of highly radioactive liquid wastes, has not yet been demonstrated on a commercial basis. Thus, no technical description can be given.

See the text for a discussion of the present state of understanding of the relative difficulty of managing the radioactive components of spent fuel under the alternatives of reprocessing and of direct transfer to final storage without reprocessing.

C.6. The costs of reprocessing

The costs associated with reprocessing have been very controversial. A wide variety of estimates has been published in the last few years. Table A-6 compares in detail the Pan Heuristics estimate of fuel cycle costs, with and without reprocessing, with several other recent evaluations. The latter were more favorable to reprocessing, but even the Pan Heuristics estimate, which found recycling clearly uneconomical in thermal reactors, was deliberately somewhat optimistic. This is evidenced by the fact that the price for reprocessing spent uranium oxide fuel recently set by Cogema of France in its contract negotiations with German utilities is reported to be $500 per kilogram (1977 dollars.)[9] This markedly exceeds the Pan Heuristics estimate of $324 per kilogram (1976 dollars). The $500 price charged by Cogema, moreover, applies to reprocessing in the new large facility with a capacity of 1600 MTU per year, planned for completion at Cap de la Hague in the late 1980s.

C.7. Fresh-fuel reprocessing

The reprocessing of fresh fuel is not a normal part of the nuclear fuel cycle. If, however, the fresh fuel contains either plutonium or highly-enriched uranium, then the reprocessing of the fresh fuel will result in the production of SSNM. Fresh fuel for LWRs which recycle plutonium (MOX fuel); fresh fuel for LMFBRs, and coprocessed fuel all contain substantial amounts of plutonium. Fresh HGTR fuel and the fresh fuel for many re-

---

[9] *Nuclear Fuel*, 2, no. 26 (26 December, 1977): 5.

Table A–6

A Comparison of Fuel Cycle Costs in Recent Evaluations of Reprocessing

Fuel Cycle Item	AGNS[a] (1976 $)	EEI[b] (1975 $)	ERDA[c] (FY 1977 $)	GESMO[d] (1975 $)	Composite: High Plutonium Value[e] (1976 $)	Composite: Low Plutonium Value[e] (1976 $)	Pan Heuristics[f] (1976 $)
Uranium Fuel Cycle Cost, per kg[g]	1341.00	920.00	1240.00	968.00	1403.00	1010.00	1002.0
Uranium, per kg U_N	136.50	81.25	110.00	79.30	136.50	87.75	70.0
Conversion to UF_6, per kg U_N	6.81	4.00	5.00	3.50	6.81	3.78	4.0
Enrichment, per SWU	94.50	75.00	100.00	75.00	94.50	81.00	108.0
Uranium Fuel Fabrication, per kg U	60.00	70.00	90.00	95.00	102.60	70.00	108.0
Direct Disposal of Spent Uranium Fuel, per kg HM	90.00	90.00	90.00	101.50	109.62	90.00	24.0[h]
Plutonium Fuel Cycle Cost, per kg HM[i]	(114.00)	210.00	793.00	278.00	(437.00)	1045.00	1775.0
Spent Fuel Transport, per kg HM	14.44	20.00	—[j]	15.00	14.44	—[j]	23.3
Reprocessing, per kg HM	149.00	115.00	280.00	150.00	124.20	263.20	324.0
Disposal of Radioactive Wastes, per kg of HM reprocessed	38.80	15.00	25.00	31.00	16.20	38.80	10.0[h]
Spent Fuel Disposal Charges Avoided, per kg of HM reprocessed	(90.00)	(90.00)	(90.00)	(101.50)	(109.20)	(90.00)	(24.0)
Value of Recovered Uranium, per kg U_R	(209.30)	(74.00)	(121.00)	(89.60)	(209.30)	(79.92)	(77.0)
Cost of Uranium Used in Plutonium Fuel	132.41	83.58	107.00	76.90	132.41	85.10	67.0
Plutonium Fuel Fabrication, per kg HM	224.00	302.00	235.00	200.00	216.00	326.16	358.6
Cost of (or Credit for) Disposing of Spent Plutonium Fuel, per kg HM	*[k]	(109.00)[l]	*[k]	(23.00)[m]	*[k]	*[k]	118.0[n]
Net Gain (or Loss) per kg of Plutonium Fuel Produced[o]	1455.00	710.00	447.00	690.00	1840.00	(35.00)	(772.0)

(Notes on following pages.)

ABBREVIATIONS for Table A-6

kg HM = kg of heavy metal (uranium, plutonium, and other transuranics).

kg U_N = kg of natural uranium metal.

kg U_R = kg of recovered uranium metal.

SWU = separative work unit.

x = final disposal charge per kg uranium spent fuel.

y = final disposal charge for wastes from reprocessing and plutonium fabrication, per kg of fuel reprocessed.

NOTES for Table A-6

General: Unless otherwise noted, figures are as reported in sources cited. Another version of this table with more extensive explanatory notes appears in Vince Taylor, "The Economics of Plutonium and Uranium," Monograph 2, ERDA Final Report, Pan Heuristics, April 30, 1977.

[a]R. J. Cholister, et al., Nuclear Fuel Cycle Closure Alternatives, Allied-General Nuclear Services (Barnwell, South Carolina), April 1976. Values used are those for the base case, summarized in Table 1, p. 3. Where costs were assumed to increase over the 20 year time period of the study, the average value was used.

[b]S. Stoller, et al., "Report on Reprocessing and Recycle of Plutonium and Uranium," Appendix VI, Nuclear Fuels Supply, Edison Electric Institute (New York, N.Y.), 30 December 1975. Data are from Table 9, p. 81 and supporting text, pp. 77-98.

[c]Benefit Analysis of Reprocessing and Recycling Light Water Reactor Fuel, U.S. Energy Research and Development Administration, ERDA 76-121, December, 1976, Table 2, p. 5, Figure 2, p. 6, and Table 3, p. 8.

[d]Final Generic Environmental Statement on the Use of Recycle Plutonium in Mixed Oxide Fuel in Light Water Cooled Reactors (GESMO), U.S. Nuclear Regulatory Commission, NUREG-0002, August 1976, Vol. 4, Chapter XI, and Table VIII-5, p. VIII-20.

[e]1975 dollar values increased by 8% to give values in 1976 dollars, and FY 1977 values decreased by 6% to give values in 1976 dollars.

fValues used here are those previously reported in Vince Taylor, Is Plutonium Really Necessary?, Pan Heuristics (Los Angeles, Ca.), presented at the Cumberland Lodge Conference on the Spread of Nuclear Weapons, 26-28 May 1976 (revised 9 September 1976, but fuel-cycle costs unchanged). Values increased by 8% to convert to 1976 dollars. No estimates of UF6 conversion, spent fuel storage or transportation, or radioactive waste transportation were included in that paper but are included here.

gFuel cycle cost is for enrichment to 3% uranium-235, the average enrichment of all replacement loads, assuming one-third BWRs and two-thirds PWRs. Production of this fuel requires 4.31 SWUs and 5.47 kg U_N, assuming enrichment plants operate at .2% tails.

hThe costs of final disposal of spent fuel and reprocessing wastes are not specified and, thus, implicitly treated as equal to zero. The figure listed is for the cost of spent-fuel temporary storage and transport (or radioactive-waste transport) to final disposal, discounted to 1 year after fuel discharge (the assumed time of spent fuel shipment to a reprocessing plant).

iFuel cycle cost is for plutonium fuel equivalent to 3% uranium-235 fuel. This requires 29 gms of fissile plutonium and .95 kg U_N. Spent fuel yields 6.1 gms of fissile plutonium per kg and approximately 1 kg of recovered uranium of 0.9% uranium-235. Allowing for the effects of uranium-236 contamination, each kg of recovered uranium is equivalent to 1.1 kg of natural uranium. In studies where different assumptions were made about the value of recovered uranium, the study assumptions were used.
To obtain the plutonium necessary for 1 kg of fuel (29 gms of fissile plutonium) requires reprocessing 4.8 kg of spent fuel. To obtain total fuel cycle cost, all items (except the uranium cost, fabrication cost, and spent fuel disposal cost or credit) are multiplied by 4.8. Credit items are indicated by parentheses.

jIncluded in reprocessing charge.

kNot reported in study, not calculated. Figures displayed for "Plutonium Fuel Cycle Cost" and "Net Gain per kg of Plutonium Fuel" implicitly treat this figure as zero.

lBased on the plutonium value of $16.92 per gm, spent MOX fuel content of 19.9 gms per kg HM plutonium, and $160 per kg HM spent fuel management charge, as reported in the study, discounted 5 years at 10% per year compounded.

[m]Calculated by author on the basis of plutonium spent fuel characteristics reported in T. H. Pigford and K. P. Ang, "The Plutonium Fuel Cycles," Health Physics (Pergamon Press), Vol. 29, October 1975, pp. 451–468, the plutonium replacement value for 1st cycle spent plutonium fuel of GESMO, Table VIII(B)–2, p. VIII(B)–3 (.66 gms uranium-235 per gm fissile Pu), and the values reported in GESMO for transportation and reprocessing of MOX spent fuel (20% higher than for spent uranium fuel). Other values were as reported in GESMO for uranium fuel.
Note: These calculations are preliminary and subject to revision.

[n]Calculated on same basis as in note m, except using Pan Heuristics values for fuel-cycle items.
Note: These calculations are preliminary and subject to revision.

[o]This equals the Uranium Fuel Cycle Cost *minus* the Plutonium Fuel Cycle Cost.

search reactors contain substantial amounts of highly enriched uranium. The same solvent extraction techniques that are used to reprocess spent fuel (described in section C.5) can be used to reprocess fresh fuel.

There are, however, two important differences between fresh-fuel reprocessing and spent-fuel reprocessing. The first is that the SSNM tends to have a considerably higher concentration in the fresh fuel than in the spent fuel. For example, the plutonium concentration in fresh MOX fuel is four times and the plutonium concentration in fresh LMFBR fuel is twenty times that of spent LWR fuel.

The second and more significant difference is that the fresh fuel is only very slightly radioactive whereas the spent fuel is highly radioactive. This difference has a number of important effects. One result is that only one solvent extraction cycle will be needed to purify the SSNM from fresh fuel whereas three are needed to purify the SSNM recovered from spent fuel. Another result is that the solvents do not degrade. This means that the contactors used for reprocessing fresh fuel can be much simpler than those used for reprocessing spent fuel and that there is no danger of a solvent explosion (as there is in the reprocessing of spent fuel).

Due to the lack of radioactivity no heavy shielding will be needed for a facility which reprocesses fresh spent fuel. Therefore, the costs of a fresh-fuel reprocessing facility are much less than those of a plant designed to reprocess spent fuel. It also means that a fresh-fuel reprocessing facility will be considerably more reliable than a spent-fuel reprocessing facility since any breakdowns that occur can be fixed directly instead of having to be done remotely.

Finally, the feasibility of fairly easy extraction of plutonium from fresh fuel underscores the observation made earlier that the danger from reprocessing lies not only in the reprocessing *plants* but in their *products*. Reprocessing done in weapon states or in multinational reprocessing centers elsewhere is not much safer than reprocessing done in national plants in nonweapon states if the international circulation of plutonium containing fresh fuel is legal.

C.8. Strategic special nuclear material, strategic special nuclear material rework, and weapon-ready nuclear material

Special nuclear material, as defined in U.S. Agreements for Nuclear Cooperation, is plutonium or uranium enriched in the isotopes uranium-235 or uranium-233. Plutonium of any isotopic composition can be used in a nuclear weapon. Not all enrichments of uranium, however, can be used in nuclear weapons. The cutoff enrichment that is usually adopted and that will be used in this report is 20 percent. Therefore, we will define strategic special nuclear material (SSNM) as being a pure compound of either plutonium or uranium with a uranium-235 or uranium-233 enrichment greater than 20 percent.

Many pure plutonium or uranium compounds are not ideally suited for use in a nuclear weapon. Plutonium nitrate, for example, which is produced at the output end of a reprocessing plant, would be difficult to use in a nuclear weapon due to its low density and liquid form. Therefore this material will have to be reworked in order to put it into a form that is appropriate for use in a nuclear weapon. The techniques for converting uranium and plutonium oxides, nitrates, and fluorides (the forms usually found in the nuclear fuel cycle) to metal are well known and fully documented in the literature,[10] as are the techniques for fabricating different uranium and plutonium metal shapes.

SSNM in metallic form which is shaped or can easily be shaped so that it can be used in a nuclear weapon we will define as being weapon-ready nuclear material (WRNM).

The amount of WRNM that is required for a nuclear weapon varies greatly depending on, among other things, the degree of neutron reflection and the weapon design. We will arbitrarily use 5 kg of fissile plutonium, 5 kg of 100 percent enriched uranium-233, and 15 kg of 100 percent enriched uranium-235 as the amount of material required for a single nuclear weapon.

2. Supplementary Tables on World Nuclear Facilities

Introduction to Power Reactor Tables

Tables A-7 and A-8 are primarily based on a listing of world nuclear power reactors in *Nuclear News*, August 1977, which contained plans as of 30 June, 1977. Some of this information has been updated by using more recent news sources. Since *Nuclear News* requires that a letter of intent be signed before a reactor is listed, it seems likely that most of the reactors in the table will eventually be built. However, the starting date listed in *Nuclear News* (and used in this table) is based on the announced reactor-construction schedule. This schedule for many reactors could easily slip by several years or more. For example, *Nuclear News's* listing of reactors scheduled to be built in the U.S. (U.S. reactors are not shown in the table) would indicate that about 150 GW of reactors would be in commercial operation by the end of 1985. However, a recent Nuclear Regulatory Commission estimate based on a reactor-by-reactor evaluation put this number around 100 GW.

Not much confidence should be placed in the precision of the numbers on plutonium production shown in the table. The actual amount of plutonium produced by each reactor will vary depending on the actual capacity factor for fueling schedule and fuel enrichment. These numbers do illustrate, however, the large quantities of plutonium that will be produced by power reactors.

---

[10] Tipton, C. R., Jr., ed., *Reactor Handbook, vol. 1, Materials*, 2d ed. (New York: Interscience Publishers, 1960); Stoller, S. M., and Richards, R. B., eds., *Reactor Handbook, vol. 2, Fuel Reprocessing*, 2d ed. (1961).

Table A-7

SUMMARY:
PRODUCTION OF PLUTONIUM IN
POWER REACTORS IN NON-WEAPON STATES*

Country	Net MWe By 1990	Separable Fissile Pu (kg) Accumulated by:				
		1970	1975	1980	1985	1990
Argentina	919	0	30	335	1,097	1,973
Austria	692	0	0	163	673	1,183
Belgium	5,475	0	0	1,131	4,284	8,556
Brazil	3,116	0	0	127	945	3,377
Bulgaria	1,760	0	0	618	1,922	3,295
Canada	15,236	51	1,400	4,706	12,304	25,340
Czechoslovakia	1,870	0	43	324	1,736	3,216
Egypt	622	0	0	0	146	631
Finland	2,160	0	0	339	1,909	3,537
German Democratic Republic	2,710	39	163	1,172	3,217	5,330
FRG	26,960	251	1,530	6,667	22,064	42,398
Hungary	1,760	0	0	0	686	2,059
India	1,684	12	345	944	2,216	3,731
Iran	9,000	0	0	0	1,872	8,049
Italy	7,295	416	803	1,500	3,049	8,508
Japan	18,708	63	1,160	7,435	19,887	33,906
Korea	1,798	0	0	185	1,147	2,658
Luxembourg	1,250	0	0	0	98	1,073
Mexico	1,308	0	0	0	829	1,792
Netherlands	503	6	131	521	911	1,301
Pakistan	125	0	48	167	286	405
Philippines	626	0	0	0	244	732
Poland	440	0	0	0	0	309
Rumania	440	0	0	0	309	652
South Africa	1,844	0	0	0	576	2,014
Spain	14,121	10	510	1,995	9,909	20,557
Sweden	9,409	0	188	3,106	9,091	16,134
Switzerland	4,021	0	525	1,527	4,068	7,101
Taiwan	4,924	0	0	355	2,762	6,468
Yugoslavia	615	0	0	96	576	1,055

*Based on August 1977 Nuclear News estimate of capacity, type and start date and 65% load factor.

Table A-8

PRODUCTION OF PLUTONIUM IN
POWER REACTORS IN NON-WEAPON STATES *

Location	Net MWe	Type	Reactor Supplier	Start Date	PuProd Kg/Yr.	Separable Fissile Pu (kg) Accumulated by				
						1970	1975	1980	1985	1990
ARGENTINA Comision Nacional de Energia Atomica										
• Atucha (Lima, Buenos Aires)	319	PHWR	Siemens	6/74	61	0	30	335	639	943
Cordoba (Embalse, Rio Tercero)	600	PHWR	AECL	1/81	114	0	0	0	458	1030
					TOTAL	0	30	335	1097	1973
AUSTRIA Gemeinschaftskernkraftwerk Tullnerfeld (GKT)										
Tullnerfeld 1 (Zwentendorf)	692	BWR	KWU/AEG	5/78	102	0	0	163	673	1183
BELGIUM Societe Belgo-Francaise d'Energie Nucleaire Mosane (SEMO)										
• Tihange 1 (Huy, Liege)	870	PWR	ACLF	9/75	136	0	0	584	1262	1941
Societe Intercommunale Belge de Gaz et d'Electricite (INTERCOM)										
Tihange 2 (Huy, Liege)	900	PWR	FRAMACECO	4/80	140	0	0	0	660	1362
Tihange 3 (Huy, Liege)	1000	PWR	ACECOWEN	10/83	156	0	0	0	187	967
Societes Reunies d'Energie du Bassin de l'Escaut (EBES)										
• Doel 1 (Antwerp)	390	PWR	ACECOWEN	2/75	61	0	0	298	602	907
• Doel 2 (Antwerp)	390	PWR	ACECOWEN	11/75	61	0	0	249	554	858
Doel 3 (Antwerp)	925	PWR	FRAMACECO	2/80	144	0	0	0	707	1429
Doel 4 (Antwerp)	1000	PWR	ACECOWEN	12/82	156	0	0	0	312	1092
					TOTAL	0	0	1131	4284	8556

*These data are based on a 1977 update and on a 65% capacity factor. This results in different values from those shown in Fig. 1, Chapter 1 in some cases. • indicates reactors that are in commercial operation.

Table A-8 (Continued)

PRODUCTION OF PLUTONIUM IN
POWER REACTORS IN NON-WEAPON STATES

Location	Net MWe	Type	Reactor Supplier	Start Date	PuProd Kg/Yr.	Separable Fissile Pu (kg) Accumulated by				
						1970	1975	1980	1985	1990
BRAZIL										
Furnas										
Angra 1 (Itaorna)	626	PWR	W	9/78	98	0	0	127	615	1104
Angra 2 (Itaorna)	1245	PWR	KWU	5/83	194	0	0	0	311	1282
Angra 3 (Itaorna)	1245	PWR	KWU	11/84	194	0	0	0	19	991
					TOTAL	0	0	127	945	3377
BULGARIA										
Kozloduy 1 (Kozloduy)	440	PWR	AEE	12/74	69	0	0	343	686	1030
Kozloduy 2 (Kozloduy)	440	PWR	AEE	12/75	69	0	0	275	618	961
Kozloduy 3 (Kozloduy)	440	PWR	AEE	80	69	0	0	0	309	652
Kozloduy 4 (Kozloduy)	440	PWR	AEE	80	69	0	0	0	309	652
					TOTAL	0	0	618	1922	3295
CANADA										
New Brunswick Electric Power Commission										
Point Lepreau (Bay of Fundy, N.B.)	600	PHWR	Can. Vic.	80	114	0	0	0	515	1087
Ontario Hydro										
Douglas Point (Tiverton, Ont.)	206	PHWR	AECL	9/68	39	51	247	444	640	837
Pickering 1 (Pickering Ont)	514	PHWR	AECL	7/71	98	0	343	833	1323	1813
Pickering 2 (Pickering Ont)	514	PHWR	AECL	12/71	98	0	294	784	1274	1764
Pickering 3 (Pickering Ont)	514	PHWR	AECL	6/72	98	0	245	735	1225	1715
Pickering 4 (Pickering Ont)	514	PHWR	AECL	6/73	98	0	147	637	1127	1617
Bruce 1 (Tiverton, Ont.)	746	PHWR	AECL	11/77	142	0	0	299	1010	1721
Bruce 2 (Tiverton, Ont.)	746	PHWR	AECL	7/77	142	0	0	356	1067	1778
Bruce 3 (Tiverton, Ont.)	746	PHWR	AECL	8/78	142	0	0	199	910	1622
Bruce 4 (Tiverton, Ont.)	746	PHWR	AECL	8/79	142	0	0	57	768	1479
Pickering 5 (Pickering Ont)	516	PHWR	AECL	4/81	98	0	0	0	364	856

Table A-8 (Continued)

PRODUCTION OF PLUTONIUM IN
POWER REACTORS IN NON-WEAPON STATES

Location	Net MWe	Type	Reactor Supplier	Start Date	PuProd Kg/Yr.	Separable Fissile Pu (kg) Accumulated by				
						1970	1975	1980	1985	1990
CANADA (cont)										
Ontario Hydro (cont)										
Pickering 6 (Pickering Ont)	516	PHWR	AECL	1/82	98	0	0	0	295	787
Pickering 7 (Pickering Ont)	516	PHWR	AECL	10/82	98	0	0	0	216	708
Pickering 8 (Pickering Ont)	516	PHWR	AECL	7/83	98	0	0	0	148	639
Bruce 5 (Tiverton, Ont.)	769	PHWR	AECL	10/83	147	0	0	0	176	909
Bruce 6 (Tiverton, Ont.)	769	PHWR	AECL	7/84	147	0	0	0	73	806
Bruce 7 (Tiverton, Ont.)	769	PHWR	AECL	4/85	147	0	0	0	0	689
Bruce 8 (Tiverton, Ont.)	769	PHWR	AECL	1/86	147	0	0	0	0	586
Darlington 1	850	PHWR	AECL	85	162	0	0	0	0	729
Darlington 2	850	PHWR	AECL	86	162	0	0	0	0	567
Darlington 3	850	PHWR	AECL	87	162	0	0	0	0	405
Darlington 4	850	PHWR	AECL	88	162	0	0	0	0	243
Hydro Quebec										
• Gentilly 1 (Becancour, Que)	250	BLWR	AECL	5/72	48	0	124	362	601	839
Gentilly 2 (Becancour, Que)	600	PHWR	AECL	12/79	114	0	0	0	572	1144
					TOTAL	51	1400	4706	12304	25340
CZECHOSLOVAKIA										
• Bohunice 1A	110	GCHWR		12/72	22	0	43	152	260	368
Bohunice 2A	440	PWR	AEE	6/78	69	0	0	103	446	789
Bohunice 2B	440	PWR	AEE	12/78	69	0	0	69	412	755
Czechoslovakian 3	440	PWR	AEE	12/79	69	0	0	0	343	686
Czechoslovakian 4	440	PWR	AEE	12/80	69	0	0	0	275	618
					TOTAL	0	43	324	1736	3216
EGYPT										
Egyptian Electricity Authority										
Sidi-Krier-1 (Sidi Krier)	622	PWR	W	83	97	0	0	0	146	631

Table A-8 (Continued)

PRODUCTION OF PLUTONIUM IN
POWER REACTORS IN NON-WEAPON STATES

Location	Net MWe	Type	Reactor Supplier	Start Date	PuProd Kg/Yr.	Separable Fissile Pu (kg) Accumulated by				
						1970	1975	1980	1985	1990
FINLAND										
Imatran Voima Osakeyhtio										
• Loviisa 1 (Loviisa)	420	PWR	AEE	5/77	66	0	0	170	498	826
Loviisa 2 (Loviisa)	420	PWR	AEE	11/78	66	0	0	72	400	727
Teollisuuden Voima Osakeyhtio										
TVO-1 (Olkiluoto)	660	BWR	ASEA-Atom	12/78	97	0	0	97	583	1070
TVO-2 (Olkiluoto)	660	BWR	ASEA-Atom	8/80	97	0	0	0	428	914
					TOTAL	0	0	339	1909	3537
GERMANY (DEMOCRATIC REPUBLIC)										
• Rheinsberg 1 (Reheinsberg, Granesee region)	70	PWR	AEE	5/66	11	39	94	149	203	258
• Nord 1 - 1(Lubmin, Greifs- wald region)	440	PWR	AEE	12/73	69	0	69	412	755	1098
• Nord 1 - 2(Lubmin, Greifs- wald region)	440	PWR	AEE	2/75	69	0	0	336	680	1023
Nord 2 - 1	440	PWR	AEE	77	69	0	0	172	515	858
Nord 2 - 2	440	PWR	AEE	78	69	0	0	103	446	789
Magdeburg 1	440	PWR	AEE	80	69	0	0	0	309	652
Magdeburg 2	440	PWR	AEE	80	69	0	0	0	309	652
					TOTAL	39	163	1172	3217	5330
GERMANY (FEDERAL REPUBLIC)										
Bayernwerk AG										
Grafenrheinfeld KKG (Grafenrheinfeld)	1225	PWR	KWU	9/79	191	0	0	57	1013	1968
Gemeinschaftkernkraftwerk Neckar (GKN)										
• GKN 1 (Necharwestheim)	805	PWR	KWU	10/76	126	0	0	402	1030	1658
GKN 2 (Necharwestheim)	805	PWR	KWU	81	126	0	0	0	440	1067

Table A-8 (Continued)

PRODUCTION OF PLUTONIUM IN
POWER REACTORS IN NON-WEAPON STATES

Location	Net MWe	Type	Reactor Supplier	Start Date	PuProd Kg/Yr.	Separable Fissile Pu (kg) Accumulated by				
						1970	1975	1980	1985	1990
GERMANY (FEDERAL REPUBLIC)--Cont.										
Kernkraftwerk Kruemmel GmbH (KKK)										
Kruemmel KKK (Geesthacht-Kruemmel/Elbe)	1260	BWR	AEG	2/80	186	0	0	0	910	1838
Hochtemperatur-Kernkraftwerk GmbH (HKG)										
THTR 300 (Hamm-Uentrop)*	300	THTR	HRB	80	neglig.	0	0	0	0	0
Kernforschungszentrum Karlsruhe										
• Karlsruhe MZFR (Karlsruhe)	52	PHWR	Siemens	10/62	10	71	121	171	220	270
Kernkraftwerk Brokdorf GmbH										
Brokdorf (Brokdorf)	1290	PWR	KWU	83	201	0	0	0	302	1308
Kernkraftwerk Brunsbuettel GmbH (KKB)										
• Brunsbuettel (Brunsbuettel/Elbe)	771	BWR	AEG	2/77	114	0	0	329	897	1465
Kernkraftwerk Hamm GmbH (KKH)										
Hamm (Hamm-Uentrop)	1300	PWR	KWU	81	203	0	0	0	710	1724
Kernkraftwerk Isar (KKI)										
Isar KKI (Ohu)	870	BWR	KWU	7/77	128	0	0	320	961	1602
Kernkraftwerk Lingen GmbH (KWL)										
• Lingen KWL (Lingen)	256	BWR	AEG	10/68	38	45	234	422	611	800

*Core contains 280 kg of 93%-enriched uranium.

Table A-8 (Continued)

PRODUCTION OF PLUTONIUM IN
POWER REACTORS IN NON-WEAPON STATES

Location	Net MWe	Type	Reactor Supplier	Start Date	PuProd Kg/Yr.	Separable Fissile Pu (kg) Accumulated by				
						1970	1975	1980	1985	1990
GERMANY (FEDERAL REPUBLIC)--Cont.										
Kernkraftwerk Obrigheim GmbH (KWO)										
• Obrigheim KWO (Obrigheim)	328	PWR	Siemens	3/69	51	41	297	553	808	1064
Kernkraftwerk Philippsburg (KKP)										
KKP 1 (Philippsburg)	864	BWR	KWU	78	127	0	0	191	827	1464
KKP 2 (Philippsburg)	1281	PWR	KWU	82	200	0	0	0	500	1499
Kernkraftwerk RWE-Bayernwerk GmbH (KRB)										
• KRB 1 Block A (Gundremmingen)	237	BWR	GE	4/67	35	94	269	443	618	793
KRB II Block C (Gundremmingen)	1249	BWR	KWU	7/81	184	0	0	0	644	1564
KRB II Block B (Gundremmingen)	1249	BWR	KWU	7/82	184	0	0	0	460	1380
Kernkraftwerk Stade GmbH (KKS)										
• Stade KKS (Stade)	630	PWR	Siemens	5/72	98	0	256	747	1238	1730
Kernkraftwerk Sud GmbH (KWS)										
Upper Rhine	1300	PWR	KWU	6/79	203	0	0	101	1115	2129
Preussiche Elektrizitats AF (PREAG)										
KKW (Nuergassen)	640	BWR	AEG	3/72	94	0	264	735	1207	1678
Kernkraftwerk Unterwesser GmbH										
• KKU (Esensham)	1230	PWR	KWU	3/77	192	0	0	537	1497	2456

Table A-8 (Continued)

PRODUCTION OF PLUTONIUM IN
POWER REACTORS IN NON-WEAPON STATES

Location	Net MWe	Type	Reactor Supplier	Start Date	PuProd Kg/Yr.	Separable Fissile Pu (kg) Accumulated by				
						1970	1975	1980	1985	1990
GERMANY (FEDERAL REPUBLIC)--Cont.										
Gemeinschaftkernkraftwerk Grohnde GmbH										
KWG (Grohnde)	1294	PWR	KWU	81	202	0	0	0	707	1716
Rheinisch-Westfalisches Elektrizitatswerk AG (RWE)										
● Biblis A (Worms/Rhein)	1146	PWR	Siemens	6/74	179	0	89	983	1877	2771
● Biblis B (Worms/Rhein)	1240	PWR	Siemens	12/76	193	0	0	580	1548	2515
● Biblis C (Worms/Rhein)	1228	PWR	KWU	82	192	0	0	0	479	1437
Kaerlich	1228	PWR	BBR	79	192	0	0	96	1054	2011
Neupotz 1 (Neupotz)	1300	PWR	BBR	83	203	0	0	0	304	1318
Neupotz 2 (Neupotz)	1300	PWR	BBR	85	203	0	0	0	0	913
Schnell-Bruter-Kernkraftwerks-gesellschaff (SBK)										
Kalkar SNR-300 (Kalkar)	282	LMFBR	Int/B-N/ NERA	82	35	0	0	0	87	260
					TOTAL	251	1530	6667	22064	42398
HUNGARY										
Hungarian Electrical Works										
Paks 1 (Paks)	440	PWR	AEE	80	69	0	0	0	309	652
Paks 2 (Paks)	440	PWR	AEE	81	69	0	0	0	240	583
Paks 3 (Paks)	440	PWR	AEE	83	69	0	0	0	103	446
Paks 4 (Paks)	440	PWR	AEE	84	69	0	0	0	34	378
					TOTAL	0	0	0	686	2059
INDIA										
Atomic Energy Commission, Department of Atomic Energy										
● Tarapur 1 (Bombay)	200	BWR	GE	10/69	29	6	153	301	448	595

Table A-8 (Continued)

PRODUCTION OF PLUTONIUM IN
POWER REACTORS IN NON-WEAPON STATES

Location	Net MWe	Type	Reactor Supplier	Start Date	PuProd Kg/Yr.	Separable Fissile Pu (kg) Accumulated by				
						1970	1975	1980	1985	1990
INDIA (Cont)										
● Tarapur 2 (Bombay)	200	BWR	GE	10/69	29	6	153	301	448	595
● RAPS 1 (Kota, Rajasthan)	202	PHWR	CGE	12/73	39	0	39	231	424	616
RAPS 2 (Kota, Rajasthan)	202	PHWR	L&T	3/78	39	0	0	69	262	454
MAPP 1 (Kalpakkam, Tamil Nadu)	220	PHWR	L&T	12/78	42	0	0	42	252	461
MAPP 2 (Kalpakkam, Tamil Nadu)	220	PHWR	L&T	6/80	42	0	0	0	189	398
Narora 1 (Uttar Pradesh)	220	PHWR		3/82	42	0	0	0	117	327
Narora 2 (Uttar Pradesh)	220	PHWR		3/83	42	0	0	0	76	285
					TOTAL	12	345	944	2216	3731
IRAN										
Atomic Energy Organization of Iran										
Iran-1 (Bushehr)	1200	PWR	KWU	80	187	0	0	0	842	1778
Iran-2 (Bushehr)	1200	PWR	KWU	81	187	0	0	0	655	1591
Iran-3	900	PWR	Fra	83	140	0	0	0	211	913
Iran-4	900	PWR	Fra	84	140	0	0	0	70	772
Not specified	1200	PWR	KWU	84	187	0	0	0	94	1030
Not specified	1200	PWR	KWU	85	187	0	0	0	0	842
Not specified	1200	PWR	KWU	86	187	0	0	0	0	655
Not specified	1200	PWR	KWU	87	187	0	0	0	0	468
					TOTAL	0	0	0	1872	8049
ITALY										
Ente Nazionale per l'Energia Elettrica (ENEL)										
● Latina (Borgo Sabotino)	150	GCR	TNPG	1/64	17	101	186	270	355	439
● Garigliano (Sessa Aurunca)	150	BWR	GE	6/64	22	122	232	343	453	564
● Trino Vercellese (Vercelli)	247	PWR	W	1/65	39	193	385	578	771	963
● Cirene (Latina)	40	LWCHW	NIRA	81	8	0	0	0	27	65

Table A-8 (Continued)

PRODUCTION OF PLUTONIUM IN
POWER REACTORS IN NON-WEAPON STATES

Location	Net MWe	Type	Reactor Supplier	Start Date	PuProd Kg/Yr.	Separable Fissile Pu (kg) Accumulated by				
						1970	1975	1980	1985	1990
ITALY (Cont)										
Caorso (Caorso, Piacenza)	840	BWR	AMN/Getsco	77	124	0	0	309	928	1547
ENEL 5 (site not yet approved)	952	PWR	EI/WENESE	84	149	0	0	0	74	817
ENEL 6 (Montalto di Castro)	982	BWR	AMN/Getsco	83	145	0	0	0	217	940
ENEL 7 (site not yet appvd)	952	PWR	EI/WENESE	84	149	0	0	0	74	817
ENEL 8 (Montalto di Castro)	982	BWR	AMN/Getsco	84	145	0	0	0	72	796
Molise #1	1000	PWR	NS	84	156	0	0	0	78	858
Molise #2	1000	PWR	NS	85	156	0	0	0	0	702
					TOTAL	416	803	1500	3049	8508
JAPAN										
Chubu Electric Power Co.										
• Hamaoka 1 (Hamaoka-cho)	516	BWR	Toshiba	3/76	76	0	0	289	669	1049
Hamaoka 2 (Hamaoka-cho)	814	BWR	Toshiba	9/78	120	0	0	156	756	1355
Hamaoka 3 (Hamaoka-cho)	1066	BWR		85	157	0	0	0	0	707
Chugoku Electric Power Co., Inc.										
• Shimane (Kashima-cho, Simane-Pref.)	439	BWR	Hitachi	3/74	65	0	52	375	699	1022
Japan Atomic Power Co. Ltd. (JAPC)										
• Tokai 1 (Tokai Mura)	159	GCR	GEC	7/66	18	63	152	242	331	421
• Tsuruga (Tsuruga)	340	BWR	GE	3/70	50	0	240	491	741	992
Tokai 2 (Tokai Mura)	1067	BWR	GE	12/77	157	0	0	314	1100	1886
Kansai Electric Power Co, Inc.										
• Mihama 1 (Mihama-cho)	320	PWR	W	11/70	50	0	205	454	704	953
• Mihama 2 (Mihama-cho)	470	PWR	MHI	7/72	73	0	183	550	917	1283
• Takahama 1 (Takahama-cho)	781	PWR	W	11/74	122	0	12	621	1231	1840

Table A-8 (Continued)

PRODUCTION OF PLUTONIUM IN
POWER REACTORS IN NON-WEAPON STATES

Location	Net MWe	Type	Reactor Supplier	Start Date	PuProd Kg/Yr.	Separable Fissile Pu (kg) Accumulated by				
						1970	1975	1980	1985	1990
JAPAN (Cont)										
● Takahama 2 (Takahama-cho)	781	PWR	MHI	11/75	122	0	0	500	1109	1718
● Mihama 3 (Mihama-cho)	781	PWR	MHI	12/76	122	0	0	366	975	1584
Ohi 1 (Ohi-cho)	1122	PWR	W	6/78	175	0	0	263	1138	2013
Ohi 2 (Ohi-cho)	1122	PWR	W	12/78	175	0	0	175	1050	1925
Kyusnu Electric Power Co. Inc.										
● Genkai 1 (Genkai, Saga)	529	PWR	MHI	10/75	83	0	0	347	759	1172
Genkai 2 (Genkai, Saga)	529	PWR	MHI	3/81	83	0	0	0	314	726
Power Reactor & Nuclear Fuel Development Corp.										
Fugen, ATR (Tsuruga)	200	LWCHW	Hitachi	6/78	42	0	0	62	270	478
Manju (Tsuruga)	300	LMFBR		1/85	36	0	0	0	0	182
Shikoku Electric Power Co.										
Ikata 1 (Ikata-cho, Ehime Pref.)	538	PWR	MHI	9/77	84	0	0	193	613	1032
Ikata 2 (Ikata-cho, Ehime Pref.)	538	PWR	MHI	10/81	84	0	0	0	269	688
Tohoku Electric Power Co., Inc.										
Onagawa (Oshikagun)	500	BWR	Toshiba	8/80	74	0	0	0	324	692
Tokyo Electric Power Co.										
● Fukushima One 1 (Fukushima)	460	BWR	GE/Toshiba	3/71	68	0	258	596	935	1274
● Fukushima One 2 (Fukushima)	784	BWR	GE	7/74	116	0	58	635	1213	1790
● Fukushima One 3 (Fukushima)	784	BWR	Toshiba	3/76	116	0	0	439	1016	1594
Fukushima One 4 (Fukushima)	784	BWR	Hitachi	10/78	116	0	0	139	716	1294
Fukushima One 5 (Fukushima)	784	BWR	Toshiba	4/78	116	0	0	196	774	1351
Fukushima One 6 (Fukushima)	1100	BWR	GE	10/79	162	0	0	32	843	1653
Fukushima Two 1 (Fukushima)	1100	BWR	Toshiba	5/82	162	0	0	0	421	1232
TOTAL						63	1160	7435	19887	33906

Table A-8 (Continued)

PRODUCTION OF PLUTONIUM IN
POWER REACTORS IN NON-WEAPON STATES

Location	Net MWe	Type	Reactor Supplier	Start Date	PuProd Kg/Yr.	Separable Fissile Pu (kg) Accumulated by				
						1970	1975	1980	1985	1990
KOREA										
Korea Electric Co.										
Ko-Ri 1 (Ko-Ri, near Pusan City)	564	PWR	W	11/77	88	0	0	185	625	1065
Ko-Ri 2 (Ko-Ri, near Pusan City)	605	PWR	W	11/82	94	0	0	0	198	670
Wolsung 1 (Near Ulsan City)	629	PHWR	AECL	4/82	120	0	0	0	324	923
					TOTAL	0	0	185	1147	2658
LUXEMBOURG										
Luxembourg Nuclear Power Company (SENU)										
Remerschen (Remerschen)	1250	PWR	BBR	7/84	195	0	0	0	98	1073
MEXICO										
Comision Federal de Electricidad										
Laguna Verde 1 (Laguna Verde, Veracruz)	654	BWR	GE	3/80	96	0	0	0	463	944
Laguna Verde 2 (Laguna Verde, Veracruz)	654	BWR	GE	3/81	96	0	0	0	366	848
					TOTAL	0	0	0	829	1792
NETHERLANDS										
Gemeenschappelijke Kernenergiecentrale Nederland NV (GKN)										
● Dodewaard (Dodewaard, Betuwe)	55	BWR	GE/GKN	3/69	8	6	47	88	128	169

Table A-8 (Continued)

PRODUCTION OF PLUTONIUM IN
POWER REACTORS IN NON-WEAPON STATES

Location	Net MWe	Type	Reactor Supplier	Start Date	PuProd Kg/Yr.	Separable Fissile Pu (kg) Accumulated by				
						1970	1975	1980	1985	1990
NETHERLANDS (Cont)										
NV Provinciale Zeeuwse Energie Maatschappij (PZEM)										
• Brossele (Borssele)	448	PWR	KWU/RDM	10/73	70	0	84	433	783	1132
					TOTAL	6	131	521	911	1301
PAKISTAN										
Pakistan Atomic Energy Commission										
• Kanupp (near Karachi)	125	PHWR	CGE	12/72	24	0	48	167	286	405
PHILIPPINES										
Philippine National Power Corp.										
Philippines 1 (Bagac, N. Luzon)	626	PWR	W	82	98	0	0	0	244	732
POLAND										
Zarnowiec (Zarnowiec)	440	PWR	AEE	85	69	0	0	0	0	309
RUMANIA										
Rumania-1 (Olt)	440	PWR	AEE	80	69	0	0	0	309	652
SOUTH AFRICA										
Electricity Supply Commission (ESCOM)										
Koeberg 1 (Koeberg)	922	PWR	Fra	82	144	0	0	0	360	1079
Koeberg 2 (Koeberg)	922	PWR	Fra	83	144	0	0	0	216	935
					TOTAL	0	0	0	576	2014

Table A-8 (Continued)

PRODUCTION OF PLUTONIUM IN
POWER REACTORS IN NON-WEAPON STATES

Location	Net MWe	Type	Reactor Supplier	Start Date	PuProd Kg/Yr.	Separable Fissile Pu (kg) Accumulated by				
						1970	1975	1980	1985	1990
SPAIN										
Central Nuclear de Asco										
Asco 1 (Asco, Tarragona)	890	PWR	W	6/79	139	0	0	69	764	1458
Asco 2 (Asco, Tarragona)	890	PWR	W	12/79	139	0	0	0	694	1388
Centrales Nucleares del Norte, SA (NUCLENOR)										
• Santa Maria de Garona (Santa Maria de Garona, Burgos)	440	BWR	GE	3/71	65	0	246	570	895	1219
Compania Sevillana de Electricidad SA, Hidroelectrica Espanola SA										
Valdecaballeros 1 (Badajoz)	937	BWR	GE	81	138	0	0	0	483	1173
Valdecaballeros 2 (Badajoz)	937	BWR	GE	82	138	0	0	0	345	1035
Compania Sevillana de Electricidad SA, Hidroelectrica Espanola SA, and Union Electrica SA										
Almaraz 1 (Almaraz, Caceres)	900	PWR	W	77	140	0	0	351	1053	1755
Almaraz 2 (Almaraz, Caceres)	900	PWR	W	78	140	0	0	211	913	1615
Electra de Viesgo										
Santillan (Santander)	970	BWR	GE	82	143	0	0	0	357	1072

Table A-8 (Continued)

PRODUCTION OF PLUTONIUM IN
POWER REACTORS IN NON-WEAPON STATES

Location	Net MWe	Type	Reactor Supplier	Start Date	PuProd Kg/Yr.	Separable Fissile Pu (kg) Accumulated by				
						1970	1975	1980	1985	1990
SPAIN (Cont)										
Electricas Reunidas de Zara-goza SA, Energia e Industrias Aragonesas SA, Union Elec-trica SA										
Trillo 1 (Trillo, Guad.)	997	PWR	KWU	6/82	156	0	0	0	389	1166
Trillo 2 (Trillo, Guad.)	997	PWR	KWU	85	156	0	0	0	0	700
Empresa Nacional Hidroelec-trica del Ribagorzana SA, and three other utilities										
Vandellos 2 (Tarragona)	900	PWR	W	12/81	140	0	0	0	421	1123
Hidroelectrica Espanola SA										
Cofrentes (Cofrentes, Valencia)	930	BWR	GE	7/80	137	0	0	0	617	1302
Cabo Cope (Cabo Cope, Murcia)	1000			Indef.						
Hispano-Francesa de Energia Nuclear, SA (HIFRENSA)										
● Vandellos (Tarragona)	480	GCR	GC	7/72	54	0	135	406	676	946
Iberduero SA										
Lemoniz 1 (Lemoniz, Vizcaya)	900	PWR	W	1/79	140	0	0	140	842	1544
Lemoniz 2 (Lemoniz, Vizcaya)	900	PWR	W	1/80	140	0	0	0	702	1404
Sayago (Moral de Sayago)	1000	PWR	W	82	156	0	0	0	390	1170
Union Electrica (UE)										
● Jose Cabrera (near Madrid)	153	PWR	W	8/69	24	10	129	248	368	487
					TOTAL	10	510	1995	9909	20557

Table A-8 (Continued)

PRODUCTION OF PLUTONIUM IN
POWER REACTORS IN NON-WEAPON STATES

Location	Net MWe	Type	Reactor Supplier	Start Date	PuProd Kg/Yr.	Separable Fissile Pu (kg) Accumulated by				
						1970	1975	1980	1985	1990
SWEDEN										
Oskarshamnsverkels Kraftgrupp AB (OKG)										
• Oskarshamn 1 (Oskarshamn)	440	BWR	ASEA-Atom	2/72	65	0	188	512	836	1160
• Oskarshamn 2 (Oskarshamn)	580	BWR	ASEA-Atom	12/74	85	0	0	427	855	1282
Oskarshamn 3 (Oskarshamn)	1060	BWR	ASEA-Atom	12/83	156	0	0	0	156	937
Statens Vattenfallsverk (SSPB)										
• Ringhals 1 (Varberg)	760	BWR	ASEA-Atom	2/76	112	0	0	437	997	1556
• Ringhals 2 (Varberg)	809	PWR	W	5/75	126	0	0	581	1212	1843
Forsmark 1 (Uppsala)	900	BWR	ASEA-Atom	7/78	133	0	0	199	862	1525
Ringhals 3 (Varberg)	900	PWR	W	12/77	140	0	0	281	983	1685
Ringhals 4 (Varberg)	900	PWR	W	7/79	140	0	0	70	772	1474
Forsmark 2 (Uppsala)	900	BWR	ASEA-Atom	7/80	133	0	0	0	597	1260
Forsmark 3 (Uppsala)	1000	BWR	ASEA-Atom	82	147	0	0	0	368	1105
Sydsvenka Kraft AB										
• Barseback 1 (Malmo)	580	BWR	ASEA-Atom	7/75	85	0	0	385	812	1239
Barseback 2 (Malmo)	580	BWR	ASEA-Atom	7/77	85	0	0	214	641	1068
					TOTAL	0	188	3106	9091	16134
SWITZERLAND										
Bernische Kraftwerke AG (BKW)										
• Muhleberg (near Berne)	306	BWR	GETSCO	10/72	45	0	99	325	550	775
Graben 1 (Graben)	1140	BWR	GETSCO	82	168	0	0	0	420	1260
Kernkraftwerk Leibstadt AG										
Leibstadt (Leibstadt)	955	BWR	GETSCO	80	141	0	0	0	633	1337

Table A-8 (Continued)

PRODUCTION OF PLUTONIUM IN
POWER REACTORS IN NON-WEAPON STATES

Location	Net MWe	Type	Reactor Supplier	Start Date	PuProd Kg/Yr.	Separable Fissile Pu (kg) Accumulated by				
						1970	1975	1980	1985	1990
SWITZERLAND (Cont)										
Nordostschweizerische Kraft-werke AG (NOK)										
● Beznau 1 (Doettingen)	350	PWR	W	12/69	55	0	273	546	819	1092
● Beznau 2 (Doettingen)	350	PWR	W	3/72	55	0	153	426	699	972
Ruethi (Ruethi)	900			Indef.						
Kernkraftwerke Goesgen-Daniken AG										
Goesgen (Daniken, SO)	920	PWR	KWU	5/78	144	0	0	230	947	1665
Kernkraftwerk Kaiseraugst AG										
Kaiseraugst (Kaiseraugst)	925	BWR	GETSCO	Indef.						
					TOTAL	0	525	1527	4068	7101
TAIWAN										
Taiwan Power Co.										
Chin-shan 1 (Shihmin Hsiang)	604	BWR	GE	77	89	0	0	222	667	1112
Chin-shan 2 (Shihmin Hsiang)	604	BWR	GE	78	89	0	0	133	578	1023
Kuosheng 1 (Kuosheng)	951	BWR	GE	4/80	140	0	0	0	659	1359
Kuosheng 2 (Kuosheng)	951	BWR	GE	4/81	140	0	0	0	518	1219
Nuclear No. 5	907	PWR	W	4/83	141	0	0	0	241	948
Nuclear No. 6	907	PWR	W	4/84	141	0	0	0	99	807
					TOTAL	0		355	2762	6468
YUGOSLAVIA										
Kisko	615	PWR	W	12/78	96	0	0	96	576	1055

Abbreviations Used in Table A-8

A-A: ASEA-Atom (Sweden)
ACECO: Association des Ateliers de Charleleroi et de Cockerill Ougree Providence (France)
ACLF group: ACECO/Creusot-Loire/ Framatome/Westinghouse Electric Energy Systems Europe (France)
ADF: Auxeltra-Delens-Francois
AECL: Atomic Energy of Canada Ltd.
AEE: Atomenergoexport (USSR) (formerly TPE: Technopromexport)
AEG: Allgemeine Elektricitaets- Gesellschaft, Aeg-Telefunken (W. Germany)
AEI: Associated Electric Industries Ltd. (U.K.)
AEPSC: American Electric Power Service Corp. (U.S.)
AETEA: Agroman, Eyt, Ea (Spain)
AGR: advanced gas-cooled reactor
Allis: Allis-Chalmers (U.S.)
Alsthom: Ste Generale de Construc- tions Electriques et Mechaniques (France)
AMN: Ansaldo Meccanico Nucleare SpA (Italy)
APC: Atomic Power Construction Ltd. (U.K.)
Arge KKU: Dyckerhoff & Widmann AG, Wayss & Freitag AG; Hegdkamp (FGR)
ASGEN: Ansaldo San Giorgio Compagnia Generale (Italy)
Aux: Auxini Ingenieria Espanola SA (Spain)
Bal: Balfour Beatty & Co. (U.K.)
BAM: Bataafsche Aanneming Maat- schappij NV (The Netherlands)
BBC: Brown Boveri et Cie (Switzer- land)
BBR: Babcock-Brown Boveri Reaktor GmbH (W. Germany)
Bech: Bechtel Corporation (U.S.)
BHE: Bharat Heavy Electrical (India)
BNDC: British Nuclear Design & Con- struction Ltd. (U.K.)
B&R: Burns & Roe, Inc. (U.S.)
B&V: Black & Veatch (U.S.)
B&W: Babcock & Wilcox Co. (U.S.)
BRAUN: C. F. Braun & Co. (U.S.)
Brown: Brown & Root, Inc. (U.S.)

BWR: boiling water reactor
Can: Canatom Ltd. (Canada)
C-B: Campenon-Bernard (France)
C-E: Combustion Engineering, Inc. (U.S.)
CEM: Compagnie Electro Mechanique (France)
CGE: Canadian General Electric
Cie GE: Cie, Generale d'Electricite (France)
CITRA: Compagnie Industrielle de Travaux (France)
CL: Creusot-Loire (France)
CNO: Construtora Noberto Oldebrecht (Brazil)
COP/TOSI/ACEC: Cockerill Ougree-Providence/ Franco Tosi SpA/Ateliers de Constructions Electriques de Charleroi SA
Daniel: Daniel Construction Co. (U.S.)
D-L: Delattre-Levivier (France)
EA: Empresarios Agrupados (Spain)
E&B: Emch & Berger (Switzerland)
Ebasco: Ebasco Services, Inc. (U.S.)
ECC: Engineering Construction Corp. (India)
EE: English Electric Co., Ltd. (U.K.)
EEC: English Electric Co. Ltd. (Canada)
EEW: English Electric and G. Wimpey Group (U.K.)
EI: Elettronucleare Italiana (Italy)
Elin: Elin Union AG (W. Germany)
ERBE: Hungarian Company for Power Plant Investment
ERDA: Energy Research and Development Administration (U.S.)
EyT: Entrecanales y Tavora SA (Spain)
FPI: Fluor Pioneer, Inc. (U.S.)
Fra: Framatome: Societe Franco-Americaine de Constructions Atomiques SA (France)
GA: General Atomic Company (U.S.)
GAAA: Groupement pour les Activites Atomiques et Avancees (France)
GC: Groupement Constructeurs Francais (France)
GCHWR: gas-cooled, heavy-water-moderated reactor
GCR: gas-cooled reactor
GE: General Electric Co. (U.S.)
GEC: General Electric Co. (U.K.)
GETSCO: General Electric Technical Ser- vices Co.
G&H: Gibbs & Hill, Inc. (U.S.)
G&HE: Gibbs and Hill Espanola SA
Gilbert: Gilbert Associates, Inc. (U.S.)
Gil/Com: Gilbert/Commonwealth (U.S.)

Abbreviations Used in Table A-8 (Continued)

GKW: Gemeinschaftkdraftwerk Weser GmbH
  (FRG)
GTM: Grands Travaux de Marseille
  (France)
Haz: Hazama Gumi Co. (Japan)
HCC: Hindustan Construction Co. (India)
Hoch: Hochtiel AG (W. Germany)
H-P: Howden-Parsons (Canada)
HQ: Hydro-Quebec (Canada)
HRB: Hochtemperatur-Reaktorbau GmbH
  (W. Germany)
HTGR: high-temperature gas-cooled
  reactor
HWLWR: heavy-water-moderated, boiling
  light-water-cooled reactor
Iber: Iberduero SA (Spain)
Imp: Impresit
INB: Internationale Natrium Brutreak-
  torbau GmbH (W. Germany)
Int: Interatom (W. Germany)
JL: John Laing Construction Ltd. (U.K.)
Jones: J. A. Jones Construction Co.
  (U.S.)
J-S: Jeumont-Schneider (France)
Kaiser: Kaiser Engineers (U.S.)
KTHTR: Konsortium THTR--Brown, Boveri
  & Cie AG, Hochtemperatur-Reaktorbau
  GmbH, Nukem GmbH (W. Germany)
Kum: Kumagaya Gumi Co. (Japan)
KWU: Kraftwerk Union AG (W. Germany)
L&T: Larsen & Toubro (India)
MAPI: Mitsubishi Atomic Power Indus-
  tries, Inc. (Japan)
Maxon: Maxon Construction Company (U.S.)
McAlp: Sir Robert McAlpine & Sons Ltd.
  (U.K.)
MECO: Montreal Engineering Co. (Canada)
MEL: Mitsubishi Electric Corporation
  (Japan)
MHI: Mitsubishi Heavy Industries, Ltd.
  (Japan)
NCC: Nuclear Civil Constructors (U.K.)
Nera: Neratoom NV (The Netherlands)
Nersa: Centrale Nucleaire Europeenne
  A Neutrons Rapides (France)
NIRA: Nucleare Italiana Reattori
  Avanzati (Italy)
NPC: Nuclear Power Co., Ltd. (U.K.)
Nuclen: Nuclebras Engenaria SA (Brazil)
NWK: Nordwestdeutsche Kraftwerke AG
  (FRG)
Obay: Obayashi Gumi Co. (Japan)

OH: Ontario Hydro (Canada)
OPS: Offshore Power Systems (U.S.)
Par (U.K.): C. A. Parsons and Co. Ltd. (U.K.)
Parsons: Ralph M. Parsons Co. (U.S.)
PE: Promon Engenharia SA (Brazil)
PHWR: pressurized heavy-water-moderated
  and -cooled reactor
PWR: pressurized water reactor
Rateau: Rateau, Ste. (France)
RDM: Rotterdamse Droogdok Madtschappij
  (The Netherlands)
RPL: Reyrolle Parsons Ltd. (U.K.)
RW: Richardsons Westgarth Ltd. (U.K.)
SACM: Societe Alsacienne de Constructions
  Mechaniques (France)
SB: Spie Batignolles SA (France)
SCG: Skanska Cementgjutereit
SEN: Sener, SA (Spain)
SGE: Societe Generale d'Enterprises (France)
SK: Sydvenska Kraft AB (Sweden)
S&L: Sargent & Lundy Engineers (U.S.)
S-L: Stal-Laval Turbin AB (Sweden)
SNC: Surveyer, Nenniger & Chenevert Inc.
  (Canada)
SO: Siemens Osterreich (Austria)
SOCIA: Societe pour l'Industrie Atomique
  (France)
SOGENE: Societa Generale per Lavore e Pub-
  bliche Utilita (Italy)
SR: Stearns-Roger Corp. (U.S.)
SS: Southern Services, Inc. (U.S.)
SSPB: Swedish State Power Board
Stork: Koninklijke Machinefabriek Stork
  (The Netherlands)
S&W: Stone & Webster Engineering Corp. (U.S.)
T&B: Townsend & Bottum, Inc.
TE: Traction-Electricite
THTR: thorium high-temperature reactor
TPC: Taiwan Power Company
TR: Tecnicas Reunidas SA (Spain)
TW: Taylor Woodrow Construction Ltd. (U.K.)
UE&C: United Engineers & Constructors (U.S.)
VBB: AB Vattenbyggnadsbyran
VMF: Verenigde Machinefabrieken NV
  (The Netherlands)
W: Westinghouse Electric Corporation (U.S.)

Introduction to Research Reactor Table

Table A-9 is based primarily on the IAEA *Directories of Nuclear Reactors*, vols. 2, 3, 5, 6, 8, and 10. The directories do not list systematically all research reactors but only those that have been described in other publications. (The IAEA is not permitted to reveal information that is obtained during the safeguarding of nuclear facilities.) The IAEA *Annual Report, 1 July 1974–30 June 1975*, pp. 78–80, which lists all research reactors that are under IAEA safeguards, was used to supplement the IAEA directories. It is the source for those research reactors which are merely listed in table A-9 and which are not characterized by precise technical information. Unfortunately, the *Annual Report* does not cover research reactors under safeguards in the Euratom countries (the Federal Republic of Germany, Italy, Belgium, and the Netherlands) nor, of course, unsafeguarded research reactors. The Egyptian WWR-C, the Indian APSARA, CIRUS, PURNIMA, and ZERLINA, and the Israeli IRR-2 are apparently the only unsafeguarded research reactors in the world and all of them are listed in the table.[11] Therefore, the table is probably complete except for the four Euratom countries.

Research reactors, being experimental facilities, are occasionally modified. Some of these modifications are permanent and some are not. Many involve changes in the power output or the fuel enrichment or both. Unfortunately, none of the sources used provides much information on these modifications. The table takes account of all of the modifications known to us.

Introduction to Breeder Reactor Table

Table A-10 lists world breeder plants. Many of these plants are quite tentative and several of the projects may never actually be carried out. Since energy independence is a major justification of the breeder reactor programs in many of the countries listed, it is very likely that these countries will insist on reprocessing and fabricating breeder reactor fuel. These activities will make easily accessible large quantities of plutonium. At a minimum, a quarter of the core inventory of each reactor will be in a readily accessible form.

---

11 *The SIPRI Yearbook* (Cambridge, Mass.: MIT Press, 1977) p. 51, purports to list all unsafeguarded nuclear facilities.

Table A-9

Operating Research Reactors in Non-Weapon States

Country	Abbreviated Name	Reactor Type	Moderator	Coolant	Power Output	Start-up Date	Fuel
Argentina	RA-0	Tank			—*		
	RA-1	Argonaut	Graphite + H_2O	H_2O	120 kW		
	RA-2	Argonaut	Graphite + H_2O	H_2O	30 kW		
	RA-3	Tank	H_2O	H_2O	5 MW	1967	90% U
	RA-4	Solid, homogeneous			--		
Australia	CF	Critical facility			--		
	HIFAR	Tank	D_2O**	D_2O	10 MW	1958	93% U
	MOATA	Argonaut	Graphite + H_2O	H_2O	10 kW	1961	90% U
Austria	ASTRA	Pool	H_2O	H_2O	5 MW	1960	90% U
	SAR-GRAZ	Argonaut	Graphite + H_2O	H_2O	1 kW	1965	20% U
	TRIGA-II-Vienna	Solid, homogeneous	$ZrH + H_2O$	H_2O	250 kW	1962	20% U
Belgium	BR-02	Tank	H_2O	H_2O	500 W	1959	90% U
	BR-1	X-10	Graphite	Air	4 MW	1956	nat. U
	BR-2	Tank	H_2O	H_2O	100 MW	1961	90% U
	BR-3/VN	Tank	$D_2O + H_2O$	D_2O+H_2O	40.9 MW	1969/72	7% U
	THETIS	Pool	H_2O	H_2O	150 kW	1967	5% U
	VENUS	Tank	$D_2O + H_2O$	D_2O+H_2O	500 W	1964	7% U
Brazil	IEAR-1	Pool	H_2O	H_2O	5 MW	1957	20% U
	TRIGA-BRAZIL	TRIGA-I	$ZrH + H_2O$***	H_2O	100 kW	1960	20% U
	REIN-1	Argonaut	Graphite + H_2O	H_2O	10 kW	1965	19.9% U

*No entry in a column means unknown. A -- entry in the power output column means unknown but < 10 kW.

**D_2O is deuterium oxide, which is heavy water.

***ZrH is zirconium hydride.

Table A-9 (Continued)

Country	Abbreviated Name	Reactor Type	Moderator	Coolant	Power Output	Start-up Date	Fuel
Bulgaria	IRT-Sofia	Pool	H_2O	H_2O	1 MW	1961	10% U
Canada	MNR	Pool	H_2O	H_2O	2 MW	1959	90% U
	NRU	Tank	D_2O	D_2O	200 MW	1957	nat. U
	NRX	Tank	D_2O	H_2O	40 MW	1947	nat. U
	PTR	Pool	H_2O	H_2O	100 W	1957	90% U
	WR-1	Tank	D_2O	Organic	60 MW	1965	2.4% U
	ZED-2	Tank	D_2O	D_2O	150 W	1960	Variable
	Slowpoke, Ottowa	Pool			20 kW		
	Slowpoke, Toronto	Pool			--		
Chile	Herald	Herald			5 MW		
Columbia	IAN-R1	Pool	H_2O	H_2O	20 kW	1965	90% U
Czechoslovakia	WWR-C-Prague	Tank	H_2O	H_2O	4 MW	1957	10% U
	SR-0	Critical			--		
	TR-0	Critical			--		
Denmark	DR-1	Aqueous homogeneous	H_2O	H_2O	2 kW	1957	20% U
	DR-2	Pool	H_2O	H_2O	5 MW	1958	90% U
	DR-3	Tank	D_2O	D_2O	10 MW	1960	93% U
Egypt	WWR-C-Cairo	Tank	H_2O	H_2O	2 MW	1961	10% U
Finland	FIR-1	Solid, homogeneous	$ZrH + H_2O$	H_2O	250 kW	1962	20% U
German Democratic Republic	WWR-S(M)	Tank	H_2O	H_2O	6 MW		10% U
	Rake II	Critical facility			--		
	RRR	Critical facility			--		

Table A-9 (Continued)

Country	Abbreviated Name	Reactor Type	Moderator	Coolant	Power Output	Start-up Date	Fuel
German Federal Republic	ADIBKA-1	Aqueous homogeneous	H_2O	H_2O	10 W	1967	93% U
	AEG NULLENERGIE REAKTOR	Tank	H_2O	H_2O	100 W	1967	Variable
	ANEX	Tank	ZrH	None	100 W	1964	Variable
	FMRB	Pool	H_2O	H_2O	1 MW	1967	90% U
	FR-2	Tank	D_2O	D_2O	44 MW	1961/66	nat. U
	FRG-1	Pool	H_2O	H_2O	5 MW	1958	20% U
	FRG-2	Pool	H_2O	H_2O	15 MW	1963	90%
	FRJ-1 (MERLIN)	Pool	H_2O	H_2O	10 MW	1962/71	80-90% U
	FRJ-2 (DIDO)	Tank	D_2O	D_2O	23 MW	1962/67	80% U
	FRM	Pool	H_2O	H_2O	4 MW	1957/68	20% U
	PR-10	Argonaut	Graphite + H_2O	H_2O	10 W	1961	20% U
	SAR-1	Argonaut	Graphite + H_2O	H_2O	10 kW	1959	20% U
	SNEAK	Fast	None	Air	1 kW	1966	20%,35%, & 93% U+Pu
	STARK	Argonaut	Graphite	H_2O	10 W	1963/64	20% U
	SUR-Aachen	Solid, homogeneous	Polyethylene	None	0.1 W	1966	20% U
	SUR-Berlin	Solid, homogeneous	Polyethylene	None	0.1 W	1963	20% U
	SUR-Bremen	Solid, homogeneous	Polyethylene	None	0.1 W	1967	20% U
	SUR-Darmstadt	Solid, homogeneous	Polyethylene	None	0.1 W	1963	20% U
	SUR-Hamburg	Solid, homogeneous	Polyethylene	None	0.1 W	1965	20% U
	SUR-Karlsruhe	Solid, homogeneous	Polyethylene	None	0.1 W	1966	20% U
	SUR-Kiel	Solid, homogeneous	Polyethylene	None	0.1 W	1966	20% U
	SUR-Munich	Solid, homogeneous	Polyethylene	None	0.1 W	1962	20% U

Table A-9 (Continued)

Country	Abbreviated Name	Reactor Type	Moderator	Coolant	Power Output	Start-up Date	Fuel
German Federal Republic (continued)	SUR-Stuttgart	Solid, homogeneous	Polyethylene	None	0.1 W	1964	20% U
	SUR-Ulm	Solid, homogeneous	Polyethylene	None	0.1 W	1965	20% U
	TRIGA-Mainz (FRMZ)	Solid, homogeneous	$ZrH + H_2O$	H_2O	100 kW	1965	20% U
	TRIGA-I-Heidelberg	Solid, homogeneous	$ZrH + H_2O$	H_2O	250 kW	1966	20% U
Greece	Democritus	Pool	H_2O	H_2O	5 MW	1961	20% U
	NTU	Sub-critical			—		
Hungary	WWR-C-Budapest	Tank	H_2O	H_2O	2 MW	1959	10% U
	ZR-4	Critical			—		
	ZR-6	Critical			—		
	Training Reactor	Tank			10 kW		
India	APSARA	Pool	H_2O	H_2O	1 MW	1956	46% U
	CIRUS	Tank	D_2O	H_2O	40 MW	1960	nat. U
	PURNIMA	Fast	None			1972	
	ZERLINA	Tank	D_2O	D_2O	400 W	1961	Variable
Indonesia	PRAB (TRIGA-II)	TRIGA-II	$ZrH + H_2O$	H_2O	1 MW	1964/71	20% U
Iran	UTRR	Pool	H_2O	H_2O	5 MW	1967	20% U
Iraq	IRT-2000	Pool	H_2O	H_2O	2 MW		10% U
Israel	IRR-1	Pool	H_2O	H_2O	5 MW	1960	90% U
	IRR-2	Tank	D_2O	D_2O	26 MW	1964	nat. U
Italy	AGN-201	Solid, homogeneous	Polyethylene	None	5 W	1960	20% U
	Avogadro RS-1	Pool	H_2O	H_2O	7 MW	1959	90% U
	ECO	Tank	D_2O	Organic	1 kW	1965	nat. U
	ESSOR	Tank	D_2O	D_2O	40 MW	1967	nat. and 90% U
	Galileo Galilei (RTS-1)	Pool	H_2O	H_2O	5 MW	1963	89.9% U
	ISPRA-1	Tank	D_2O	D_2O	5 MW	1959	20% U

Table A-9 (Continued)

Country	Abbreviated Name	Reactor Type	Moderator	Coolant	Power Output	Start-up Date	Fuel
Italy (continued)	L-54	Aqueous homogeneous	H_2O	H_2O	50 kW	1960	20% U
	RANA	Pool	H_2O	H_2O	10 kW	1961/65	20% U
	RB-1		Graphite	None	10 W	1962	20% U
	RB-2	Argonaut	Graphite + H_2O	H_2O	10 kW	1963	20% U
	RC-1	Solid, homogeneous	ZrH + H_2O	H_2O	1 MW	1960	20% U
	RITMO (RC-4)	Pool	H_2O	H_2O	100 W	1965	90% U
	ROSPO	Tank	Organic	None	neg.	1963	89.9% U
	TAPIRO	Fast	None	Helium	5 kW	1971	93.5% U
	TRIGA-II-PAVIA (LENA)	Solid, homogeneous	ZrH + H_2O	H_2O	250 kW	1965	20% U
Japan	AHCF	Aqueous homogeneous	D_2O	D_2O	50 W	1961	20% U
	DCA	Critical facility			--		
	FCA	Critical facility			10 kW		
	HTR	Pool	H_2O	H_2O	100 kW	1961	10% U
	JMTR	Tank	H_2O	H_2O	50 MW	1968	90% U
	JMTRC	Pool	H_2O	H_2O	10 W	1965/67	90% U
	JRR-2	Tank	D_2O	D_2O	10 MW	1960	90% U
	JRR-3	Tank	D_2O	D_2O	10 MW	1962	nat. U
	JRR-4	Pool	H_2O	H_2O	1 MW	1965	90% U
	KUR	Pool	H_2O	H_2O	5 MW	1964	90% U
	KUCA	Critical facility			--		
	NCA	Tank	H_2O	H_2O	200 W	1963	1-3% U
	NSRR	Triga, Pulse	ZrH + H_2O	H_2O	300 kW		
	OCF	Tank	H_2O	H_2O	100 W	1962	1.5-2.5% U

Table A-9 (Continued)

Country	Abbreviated Name	Reactor Type	Moderator	Coolant	Power Output	Start-up Date	Fuel
Japan (continued)	SHCA	Solid, homogeneous	Graphite	None	10 W	1961	20% U
	TCA	Tank	H_2O	H_2O	200 W	1962	2.6% U
	TODAI	Fast	None		2 kW		
	TRIGA-II-Musashi	Solid, homogeneous	$ZrH + H_2O$	H_2O	100 kW	1963	20% U
	TRIGA-II-Rikkyo	Solid, homogeneous	$ZrH + H_2O$	H_2O	100 kW	1961	20% U
	TTR-1	Pool	H_2O	H_2O	100 kW	1962	20% U
	UTR-10-Kinki	Argonaut	Graphite + H_2O	H_2O	0.1 W	1961	90% U
Korea	KRR-TRIGA-III	TRIGA-III	$ZrH + H_2O$	H_2O	2 MW		
	TRIGA-II-SEOUL	Solid, homogeneous	$ZrH + H_2O$	H_2O	250 kW	1962	20% U
Libya	Unknown				10 MW		
Mexico	Centro Nuclear	TRIGA-III	$ZrH + H_2O$	H_2O	1 MW		
	Training Reactor Facility	SUR-100			--		
Netherlands	HFR	Tank	H_2O	H_2O	30 MW	1961	90% U
	HOR	Pool	H_2O	H_2O	2 MW	1963	90% U
	LFR	Argonaut	Graphite + H_2O	H_2O	10 kW	1960	90% U
	SUSPOP	Aqueous homogeneous	H_2O	H_2O	neg.	1959	20% U
Norway	HBWR	Tank	D_2O	D_2O	20 MW	1959	1.5% U
	JEEP-2	Tank	D_2O	D_2O	2 MW	1966	2.5% U
	NORA	Tank	$D_2O + H_2O$	$D_2O + H_2O$	100 W	1961	nat. U
Pakistan	PARR	Pool	H_2O	H_2O	5 MW	1965	90% U
Philippines	PRR-1	Pool	H_2O	H_2O	1 MW	1963	20% U

Table A-9 (Continued)

Country	Abbreviated Name	Reactor Type	Moderator	Coolant	Power Output	Start-up Date	Fuel
Poland	AGATA	Critical facility			--		
	ANNA	Critical facility	Graphite + H_2O	H_2O	100 W	1963	21% U
	MARIA	Tank	H_2O	H_2O	30 MW	1967	10-36% U
	MARYLA	Pool	H_2O	H_2O	10 kW	1958	10% U
	WWR-C-Warsaw (EVA)	Tank	H_2O	H_2O	10 MW	1961	20% U
Portugal	RPI	Pool			1 MW		
Rumania	TRIGA	Solid, homogeneous	ZrH + H_2O	H_2O	5 MW	1957	93% U
	WWR-C – Bucharest	Tank	H_2O	H_2O	3 MW	1965	10% U
South Africa	SAFARI-1	Tank	H_2O	H_2O	20 MW	1962	90% U
Spain	ARBI	Argonaut	Graphite + H_2O	H_2O	10 kW	1960	20% U
	ARGOS	Argonaut	Graphite + H_2O	H_2O	10 kW		20% U
	CORAL-1	Fast	None	None	10 W	1968	90% U
	JEN-1	Pool	H_2O	H_2O	3 MW	1958	20% U
	JEN-2	Pool			--		
Sweden	KRITZ	Critical facility			--		
	R-1	Tank	D_2O	D_2O	1 MW	1964	nat. U
	R-2	Tank	H_2O	H_2O	50 MW	1960	90% U
	R2-0	Pool	H_2O	H_2O	1 MW	1961	90% U
Switzerland	AGN-201P	Solid, homogeneous	Polyethylene	None	20 W	1958	20% U
	AGN-211P	Solid, homogeneous	Polyethylene	None	2 kW	1961	20% U
	CROCUS	Critical facility			--		
	DIORIT	Tank	D_2O	D_2O	30 MW	1960	nat. U
	PROTEUS	Critical facility	Graphite	None	5 kW	1968	5% U
	SAPHIR	Pool	H_2O	H_2O	5 MW	1957	20% U

Table A-9 (Continued)

Country	Abbreviated Name	Reactor Type	Moderator	Coolant	Power Output	Start-up Date	Fuel
Taiwan	THOR	Pool	H_2O	H_2O	1 MW	1961	20% U
	TRR	NRX	D_2O	H_2O	40 MW	1973	nat. U
	ZPRL	Pool			10 kW		
	THAR	Argonaut	Graphite + H_2O	H_2O	10 kW		
	MER	Mobile educational			--		
Thailand	TRR-1	Pool	H_2O	H_2O	1 MW	1962	90% U
Turkey	TR-1	Pool	H_2O	H_2O	1 MW	1962	90% U
Uruguay	RUDI	Lockheed			100 kW		
Venezuela	RV-1	Pool	H_2O	H_2O	3 MW	1960/65	20% U
Yugoslavia	R-A	Tank	D_2O	D_2O	10 MW	1959	2% U
	R-B	Tank	D_2O	D_2O	neg.	1958	nat. U
	TRIGA-II-Ljubljana	Solid, homogeneous	$ZrH + H_2O$	H_2O	250 kW	1966	20% U
Zaire	TRICO	Solid, homogeneous	$ZrH + H_2O$	H_2O	1 MW	1959/72	20% U

Table A-10

Breeder Projects--Non-Weapons States

Country, Name of Reactor (Location)	Type of Reactor	Status	Power Capacity	Year of Criticality	Comments
Brazil					
Cobra	Experimental fast breeder power reactor	Planned			Part of agreement signed with France 4 July 1975; design expected to be similar to Rapsodie and Phenix*
Germany, FR					
SNEAK (Schnelle Null-Energie Anordnung Karlsruhe)	Fast research reactor	In operation		1966	Used to investigate the neutron physics of large fast breeder reactors
KNK II (Karlsruhe)	Fast research reactor	In operation	19 MWe	1977	Converted from sodium-cooled research reactor KNK I (start-up date 1971)
SNR-300 (Kalkar)	Liquid-metal fast breeder power reactor	Under construction	282 MWe	1982	Prototype-scale LMFBR; construction, in cooperation with Belgium and the Netherlands. Core to contain over 1000 kg of plutonium
SNR-2 (Kalkar)	Liquid-metal fast breeder power reactor	Planned	1200 MWe	Indef.	Commercial LMFBR; construction expected to begin after completion of SNR-300 in mid-1980s
India					
Purnima	Fast research reactor	In operation		1972	Used to provide data on the use of plutonium in fast breeder reactors
FBTR (Fast Breeder Test Reactor) (Kalpakkam)	Experimental fast breeder power reactor	Under construction	15 MWe	1980	Design basically similar to Rapsodie* but modified for power generation; French Atomic Energy Commission assisting in design and construction; will in particular be used in research on the breeding of fissile material in thorium

*The fast research reactor Rapsodie (40 MWt) and the prototype LMFBR Phenix (233 MWe) are part of the French breeder program. Respectively, they became critical in 1967 and 1973. The Rapsodie core contains 40 kg of plutonium and 94 kg of highly-enriched uranium (85%). The Phenix core contains 840 kg of plutonium.

Table A-10 (Continued)

Country, Name of Reactor (Location)	Type of Reactor	Status	Power Capacity	Year of Criticality	Comments
Iraq Unnamed (not selected)	Liquid-metal fast breeder power reactor	Planned			Bilateral agreement with France signed in November 1975 includes the eventual construction of an LMFBR similar to the Phenix, after construction of a PWR plant
Italy Tapiro	Fast research reactor	In operation	5 kWt	1971	Neutron physics and radiobiology; uses 23 kg highly enriched (93.5%) uranium
PEC (Brasimone)	Fast research reactor	Under construction	120 MWt	1980	Design similar to FFTF* (USA) but smaller; delays in starting project due to disagreements within Euratom
Japan Joyo (Oarai)	Liquid-metal fast breeder research reactor	In operation	50 MWt	1977	Design and construction by Power Reactor Nuclear Fuel Development Corporation (PNC) Core contains 150 kg of plutonium
Monju (Monju)	Liquid-metal fast breeder power reactor	Planned	300 MWe	1985	Prototype-scale LMFBR under development by PNC; design work reported at an advanced stage. Core to contain over 1000 kg of plutonium
Spain Coral-1	Fast research reactor	In operation	10 MWt	1968	Fast-neutron source and fast-neutron physics studies; uses 22 kg of highly-enriched (90%) uranium

SOURCES: World Armaments and Disarmaments SIPRI Yearbook 1976; Nuclear Engineering International, April, September, and October 1977, and July 1976; Nuclear News, April, 1977; and An Assessment of International Policies on Breeder Reactors, December 1977, Appendix F, prepared by International Energy Associates Ltd.; IAEA Directory of Nuclear Reactors, Vol. IX.

*The Fast Flux Test Facility (FFTF) is a 400 MWt fast research reactor. It is scheduled to become critical in 1978 and will contain 600 kg of plutonium in its core.

Table A-11

Non-Communist Uranium Enrichment Facilities

Country/Name of Plant	Type	Capacity 10^6 SWU	Schedule
Brazil			
Resende (Nuclebras)	Nozzle	0.18	1979
Resende (Nuclebras)	Nozzle	2	1989
France			
Pierrelatte (Military)	Diffusion	0.4-0.5	1966
Tricastin (Eurodif)	Diffusion	10.8 (2.5)	1978-81 (1979)
Tricastin (Coredif)	Diffusion	9-10	1985
Germany			
Karlsruhe Nuclear Center	Nozzle	0.002	1967 (shutdown in 1972)
Almelo, Netherlands (Urenco)	Centrifuge	0.025	1974
Japan			
PNC Ningyotoge	Centrifuge	0.01	1981
PNC	Centrifuge	2	1988
Netherlands			
Almelo (Urenco)	Centrifuge	0.025	1974
Almelo (Urenco)	Centrifuge	0.2 (0.06)	1978 (1977)
Almelo (Urenco)	Centrifuge	0.8 add-on	1983/84
South Africa			
UCOR (Valindaba)	Advanced Vortex Tube	0.006	1977
UCOR (Valindaba)	Advanced Vortex Tube	5*	1986/87
United Kingdom			
Capenhurst (Military)	Diffusion	0.4	1953
Capenhurst (Urenco)	Centrifuge	0.025	1974
Capenhurst (Urenco)	Centrifuge	0.2 (0.05)	1979 (1977)
Capenhurst (Urenco)	Centrifuge	0.8 add-on	1983/84
United States			
Oak Ridge	Diffusion	4.73**	1951
Paducah	Diffusion	7.31**	1954
Portsmouth	Diffusion	5.19**	1956
Portsmouth	Centrifuge	8.8	1986 (full capacity: 1988)

*Current estimate; capacity to be determined at the end of 1978.

**CIP (Cascade Improvement Program) and CUP (Cascade Upgrade Program) will increase the capacity of the three diffusion plants to a total of 27.7 million SWU by 1981.

Table A-12

MAJOR NONCOMMUNIST WORLD REPROCESSING FACILITIES

Country	Facility	Fuel Type	Design Capacity	Status
Argentina	Ezeiza	Metal (research reactor fuel)	Lab-scale	Shut-down
Belgium	Eurochemic-Mol	Multipurpose plant	60 te/yr	Start-up 1966; shut-down 1974. Has been used for reprocessing development. Plans to restart, possibly by 1981/82.
Brazil		UO_2		Planned as part of package deal with Federal Republic of Germany
France	La Hague (UP2)	Natural U metal	800 te/yr	Start-up in 1966. Oxide head-end being installed which will permit reprocessing of UO_2 fuel. 15 te of oxide fuel reprocessed in 1976. A further test of 60 te of oxide fuel was planned for 1977 but only 10 te were actually reprocessed. Capacity is scheduled to increase to 400 te/yr by 1979 and 800 te/yr in 1981.
	La Hague (UP3a) (UP3b)	UO_2	800 te/yr 800 te/yr	1986 start-up; for foreign demand 1987 start-up; primarily for domestic use
	Marcoule (UP1)	Natural U metal	900–1200 te/yr	Start-up 1958 for military purposes; will take over commercial natural U metal reprocessing from La Hague.
Federal Republic of Germany	KEWA (United Reprocessors)	UO_2	1500 te/yr	Start-up in late 1980s
	WAK Karlsruhe	Breeder and UO_2	40 te/yr pilot plant	In operation since 1970
India	Trombay	Metal	100 te/yr	In operation since 1965
	Tarapur	Metal and UO_2	150 te/yr	Being cold tested

TableA-12(Continued)

Country	Facility	Fuel Type	Design Capacity	Status
Italy	Eurex-1-Sallugia	UO_2 and metal	10 te/yr	In operation
	Unnamed	UO_2	500 te/yr	Start-up 1985; plans temporarily shelved
Japan	PNC Tokai-Mura	UO_2	210 te/yr	1977 Hot testing started after U.S./Japanese Agreement for it to operate as an experimental facility
	Unnamed	UO_2	1000 te/yr	Start-up in late 1980s if site can be found
Pakistan	Unnamed	UO_2	unknown	Planned for construction under an agreement with France
Spain	Juan Vigon Ctr., Madrid	Metal	Small pilot plant	Shut down
	Unnamed	UO_2	1000 te/yr	Possibly by mid-1980s
United Kingdom	British Nuclear Fuels, Ltd., Windscale	Natural U metal	2400 te/yr	In operation since 1964. Installed oxide head-end in 1972 with 300 te/yr capacity but was shut down after an incident in 1973. Refurbished oxide head-end with 200–400 te/yr capacity may be installed in 1980
	British Nuclear Fuels, Ltd.	UO_2	1000 te/yr	Start-up planned for 1987; half for expected domestic requirements. Plans pending during ongoing hearings on the facility
United States	NFS West Valley, New York	UO_2	300 te/yr	Operated from 1966 to 1972; permanently shut down
	GE Morris, Illinois	UO_2	300 te/yr	Inoperable in present form; very likely will never open
	AGNS Barnwell, S.C.	UO_2	1500 te/yr	Earliest possible start is 1982
Yugoslavia	Boris Kidric Institute	Metal	Lab-scale	Shut down

Table A-12

MAJOR NONCOMMUNIST WORLD REPROCESSING FACILITIES

SOURCES: Library of Congress, Congressional Research Service, Facts on Nuclear Proliferation, U.S. Senate Committee on Government Operations (U.S. Government Printing Office, Washington, D.C.), December 1975; "Reports on Nuclear Programs around the World," Nuclear News, mid-February 1976, pp. 47–51, and August 1977, p. 56; Nuc. Eng. Internat., February 1976, pp. 21–27, April/May 1976, pp. 20–21, and July 1977, p. 36; Nucleonics Weekly, December 27, 1977, p. 6; and Nuclear Fuel, September, 19, 1977, p. 5.

Table A-13

Summary of Heavy Water Separation Plants

Site, Constructor, Operator	Nominal Capacity (Mg/yr)	Year of Start Up	Status	Boundary Conditions	Source	Extraction	Enrichment Initial	Final
Rjukan, Norway Norsk Hydro	20	1934	Operating	PA	HY	(a)	WE	WD
Savannah River, USA Girdler, Lummus, Dupont	190	1952 (b)	Operating	SC	RW	GS	GS DT	WD WE
Nangal, India Linde, DAE	14	1962	?	PA	RW	WE	WE	HD
Port Hawkesbury, Canada, Lummus, CGE, AECL	360	1970	Operating	SC	RW	GS	GS DT	WD
Bruce A, Canada Lummus, Ontario Hyd	720	1973	Operating	SC	RW	GS	GS DT	WD
Glace Bay, Canada Canatom, AECL	360	1976	Constr.	SC	RW	GS	GS DT	WD
Barooda, India GELPRA, DAE	67	Indef.[1]	Constr.	PA	SG	AH	AH MO	AH MO
Kota, India DAE	100	1979[2]	Constr.	SC	RW	GS	GS DT	WD
Tuticorin, India GELPRA, DAE	71	1978[2]	Constr.	PA	SG	AH	AH MO	AH MO
Talcher, India Uhde, DAE	63	1979[2]	Constr.	PA	SG	AH	AH DT	WD
Bruce B, Canada Lummus, Ont. Hydro	720	1978	Constr.	SC	RW	GS	GS DT	WD
La Prade, Canada Canatom, AECL	720	1982	Constr.	SC	RW	GS	GS DT	WD

Legend:

SC	Self-contained	WE	Water electrolysis
PA	Parasitic	WD	Water distillation
RW	River water	AD	Ammonia distillation
HY	Hydrogen	HD	Hydrogen distillation
SG	Ammonia synthesis gas	GC	Gas phase heterogeneous catalyst
GS	Girdler sulfide	DT	Dual temperature enrichment
AH	Ammonia-hydrogen exchange	MO	Monothermal enrichment

AECL   Atomic Energy of Canada Ltd.
CCM   Companie de Construction Mécanique Sulzer
CEA   Commisariat à l'Energie Atomique
CGE   Canadian General Electric Ltd.

DAE   Department of Atomic Energy (India)
GELPRA   Groupement Eau Lourde Procédé Ammoniac
SAL   Société à l'Air Liquide
SCC   Société Chimique de Charbonnages

(a)   Initially WE, later GC and WD added.
(b)   Original capacity about 500 Mg/yr. Two thirds shutdown and recently dismantled.

[1] Start up indefinitely delayed due to an explosion during testing, December 1977.

[2] Year shown modified to show probable slippage since table originally prepared.
Otherwise table drawn from *Nuclear Engineering International*, September 1976, p. 11.

APPENDIX B:
Methods for Estimating the Viability of Medium-Sized Silo-Based and Submarine-Based Forces

This appendix documents the means by which the effectiveness of attacks against silo-based and submarine-based forces presented in chapter 5 were calculated. In the case of silo-based forces, three types of measures are presented: the expected number of surviving silos from a given attack against a given number of silos; the probability that none or no more than one silo would survive; and the number of silos with ICBMs that would have to be deployed in order to have a given expected number survive an attack of a given size.

The essence of the attacks on the submarine-based systems is to wait until conditions are right before attacking. Hence, the type of measure presented is the probability that the criteria for attack will be met within a given interval of time (that is, that at least K submarines will be simultaneously under trail at least once in an interval of time).

Silo-Based Missile Formulas and Calculations

Let R be the distance in nautical miles from ground-zero at which a given static overpressure, H, is expected to occur from a groundburst weapon of yield Y. Then to a good approximation

$$R = Y^{1/3} \cdot \{[-1 + (1 + 0.36H)^{1/2}]/2.3\}^{-2/3}$$

if H is expressed in units of pounds per square inch and Y in megatons. This formula is derived from an approximation by Brode.[1] Since the formula is for groundburst weapons it gives less than the maximum radius possible from an airburst weapon, but the difference is unimportant for high overpressures.

For a point target of nominal hardness H the probability of surviving a reliably delivered weapon is q where

$$q = (0.5)^x$$

$$x = R^2/[(CEP)^2 + R^2 \cdot \sigma^2 \cdot \ln(4)],$$

CEP is the circular error probable, and σ is a dimensionless measure of the variance of the probability of damage as a function of distance between actual ground zero and target[2]. For a "cookie cutter" damage function, $\sigma = 0$. We use a value $\sigma = 0.2$.[3]

---

[1] Brode, H., "Review of Nuclear Weapons Effects," *Annual Review of Nuclear Sciences* 18 (1968): 180.

[2] The blast resistance of a target generally decreases with increasing yield. Thus a target that has a nominal hardness of about 500 psi against weapons in the 20–40 kiloton range might have a nominal hardness of only 300 psi against megaton-range weapons. Hence, strictly speaking, when a target hardness is specified, a yield range for which it is applicable should also be given.

[3] For a discussion of these approximations, see *Mathematical Background and Programming Aids for the Physical Vulnerability System for Nuclear Weapons*, DI-550-27-4, 1

We assume that fratricide considerations lead to two time windows per target;[4] and that if more than one weapon passes through a time window, only one is effective. If n_1 weapons are allocated to one time window and n_2 to the other, the probability that at least one reliable weapon arrives at the first is r_1:

$$r_1 = 1 - (1 - R)^{n_1}$$

where R is the overall reliability per warhead. Similarly,

$$r_2 = 1 - (1 - R)^{n_2}$$

Since at most one weapon can be effective per time window, the overall survival probability of a target, s, is

$$s = [1 - r_1(1 - q)] \cdot [1 - r_2(1 - q)].$$

For T targets, each with the same value of s, the expected (average) number of survivors is Ts. The probability that none will survive is just the probability that all are killed, which is $(1 - s)^T$. Similarly, using the formulas of the binomial distribution, the probability that exactly one will survive is $T \cdot s \cdot (1 - s)^{T-1}$.

For a target hardened to 300 psi against a 5-megaton weapon, $R = 0.67$ nautical mile. For a *CEP* of one-quarter nautical mile, then $x = 5.1$ and hence $q = 0.029$. The following table gives values of the survival probability per target for various numbers of these weapons per target, each with reliability 0.75.

Weapons per Target	Survival Probability per Target
1	0.272
2	0.074
3	0.024
4	0.008

An example is the best way to illustrate how to find the number of silos with missiles which must be deployed in order that the expected number of survivors from a specified attack be equal to a specified value. Suppose that the objective is to deploy enough silos with missiles so that the expected number that would survive a 500 warhead attack would be 40. (The attacking warheads are assumed to be of the type discussed just above.)

First consider cases where the number of weapons per silo is an integer and find two such cases that bracket the desired result. In the case at hand, if there are 500 silos (one warhead per silo) the expected number of survivors is $0.272(500) = 136$. If there are 250 silos (two warheads per silo) the expected number of survivors is $0.074(250) = 18.5$. The correct

---

November 1974, Defense Intelligence Agency. The formula for q given above results if the circular coverage function is used to approximate the damage function, as was the case for the DIA system until 1972. It remains valid as an approximation for $\sigma \leq 0.30$, as discussed on p. 5 of the reference.

[4] See the main text for a discussion of fratricide and time windows.

number of silos thus lies between 250 and 500. Assume S silos are deployed and $250 \leq S \leq 500$. Then $500 - S$ silos will be allocated two warheads and the remainder, $2S - 500$, will be allocated one warhead. The expected number of survivors will be $0.074(500 - S) + 0.272(2S - 500)$. Setting this expression equal to 40 and solving for S yields $S = 295.74$, or rounding, 296.

If the attack size is increased to, say, 1500 warheads, then a few trials will show that the correct number of silos lies between enough to dilute the attack to two weapons per target (750 silos) and enough to dilute it to three weapons per target (500 silos). Reasoning as above, the answer is found by solving for S in the equation $0.024(1500 - 2S) + 0.074(3S - 1500) = 40$. With rounding, $S = 661$.

Submarine Search and Trail Models

Let m be the number of SSNs (or other ASW search units) that are searching an area A. Let s be the sweep rate per unit.

We can reasonably expect individual search units to be sufficiently coordinated that they will not have overlapping search areas; hence, their search rates can be assumed additive. We assume that the conditional probability an SSBN will be detected in a small time interval dt is independent of time, and is equal to the fraction of the total area swept in the interval dt, $(ms/A)dt$.

Consider the case of one SSBN. Let $P(t)$ be the probability that it escapes detection from time 0 to t. Then

$$P(t + dt) = P(t) \cdot [1 - (ms/A)dt];$$

that is, the probability of escaping detection by time t times the conditional probability of escaping detection in the time interval dt is just the probability of escaping detection by time $t + dt$. From this, $P'(t) = -(ms/A)P(t)$. Since $P(0) = 1$ (no chance of detection before the starting time, $t = 0$), this differential equation has the solution

$$P(t) = e^{-(ms/A)t}, t \geq 0.$$

Consider now the case of n SSBNs. Suppose, at a particular time t, the number of SSBNs under trail is j, $0 \leq j \leq n$. We assume that each SSBN under trail occupies one search unit, so that the number actually searching is $m - j$. Since $n - j$ SSBNs are to be found, the probability of finding another SSBN in time interval dt—and hence going from j to $j + 1$ SSBNs under trail—is:

$$P(j \rightarrow j + 1,\ dt) = (m - j)(n - j)\frac{s}{A}\,dt,\ 0 \leq j \leq n - 1.$$

Similarly, if H is the mean holding time, the probability of losing an SSBN under trail in time dt and hence going from j to $j - 1$ is

$$P(j \rightarrow j - 1,\ dt) = j\left(\frac{1}{H}\right)\,dt,\ 1 \leq j \leq n.$$

These are the transition probabilities for a Markov process.[5] The probability that this process is in state j at time t is the probability that j submarines would be under trail at time t. However, what we want is the probability that at least some threshold number, K, of submarines has been under trail at least once in the interval from time 0 to t.

We can modify the process to compute such a probability by making the state $j = K$ a "capture state" so that once the modified process enters state K it never leaves it. Then the probability that the modified process is in state K at time t is just the cumulative probability that the seeking force will have located K submarines (and presumably attacked them) at least once at or before time t.

The modified process has only the $K + 1$ states $j = 0, \ldots, K$. Within this restricted range the transition probabilities are identical to those above, except that $P(K \rightarrow K - 1, dt) = 0$. The vector of probabilities that the process is in state j, $0 \le j \le K$, is a solution to a system of first order linear differential equations with constant coefficients[6] and thus readily amenable to calculation by computer.

[5] In fact, a rather simple "birth and death" process with only a finite number of states. See W. Feller, *An Introduction to Probability Theory and its Applications*, vol. 1, 3rd ed. (New York: Wiley and Sons, 1968) especially pp. 454 ff.

[6] *Ibid.*

Index